Mastering the TI-92

Explorations from Algebra through Calculus

Nelson Rich
Department of Mathematics and Computer Science
Nazareth College
Rochester, NY

Judith Rose
Department of Mathematics and Computer Science
Nazareth College
Rochester, NY

Lawrence Gilligan
Department of Mathematics, Physics, and Computing Technology
OMI College of Applied Science
University of Cincinnati

GILMAR Publishing
P.O. Box 6376
Cincinnati, OH 45206
(513) 751-8688

Printed in the United States of America.

International Standard Book Number: **0-9626661-9-x**

2 3 4 5 6 7 8 9 0

First Printing: February, 1996
Second Printing: August, 1996

The cover was designed by Patricia K. Lloyd.

GILMAR Publishing Company
P.O. Box 6376
Cincinnati, OH 45206
(513) 751-8688   FAX: (513) 751-2821

CONTENTS

PART 2: Explorations into the Mathematics Curricula

PART 3: Programming the TI-92 177

Other Books by the Authors:

The TI-85 Reference Guide, Rich/Gilligan, GILMAR Publishing (Cincinnati: 1993).

Applied Calculus, Fourth Edition, Taylor/Gilligan, Brooks/Cole Publishing Company (Pacific Grove, CA: 1996)

Calculus and the DERIVE® Program: Experiments with the Computer, Third Edition, Gilligan/Marquardt, GILMAR Publishing (Cincinnati: 1995)

Linear Algebra Experiments Using DERIVE®, GILMAR Publishing, Salter/Gilligan (Cincinnati: 1992)

Precalculus Experiments with the TI-81 Graphics Calculator, Gilligan, D. C. Heath and Company (Lexington, MA: 1991)

Precalculus Experiments with the Casio Graphics Calculator, Gilligan, D. C. Heath and Company (Lexington, MA: 1991)

Preface

This manual was written to try to provide assistance to new users of the *TI-92*. It is meant to *supplement* not *replace* the *TI-92 Guidebook*.

One of the strengths of this book is that it provides over 600 screens of the *TI-92*'s various applications. It also provides hundreds of examples demonstrating how to use the various keys, menus, and submenus.

The contents are separated into three parts: Part One is meant as an overview of virtually all of the important features of the *TI-92*. It is intended to provide the reader with a certain level of "comfort" with the *TI-92* as well as a continuing reference for the machine's features. Part Two takes the reader into the mathematics curriculum by applying the *TI-92* in twelve *Explorations*. These explorations (or projects) are in the areas of algebra, geometry, precalculus, and calculus. Part Three highlights the programming features of the *TI-92*; we have included eleven programs of varying degrees of difficulty.

We welcome comments and suggestions from readers. You may find it most convenient to communicate via electronic mail:

Nelson Rich	rich@naz.edu	
Judith Rose	jxrose@naz.edu	
Lawrence Gilligan	gilligan@uc.edu	http://www.uc.edu/~gilligan

or write to us c/o Gilmar Publishing, P.O. Box 6376, Cincinnati, OH 45206.

Acknowledgments: The staff at Texas Instruments has been extremely cooperative and encouraging during the life of this project. We would especially like to express our gratitude to Jeffrey Crump for his assistance. Also, Jerry and Joyce Glynn, MathWare, provided a critique of the manuscript as well as an inordinate amount of enthusiasm which helped to get us through this project. Thanks Jerry and Joyce.

Patricia Lloyd, University of Cincinnati, designed the cover and, once again, we appreciate her expertise.

Dedication

For Charlotte, thank you for everything.
For Florence, pleasant dreams.

N.R.

To the late Edwin J. Purcell, who instilled in me a love for the language of mathematics.

J.R.

For Andy and Katie, who continue to make each day challenging but so very rewarding.

L.G.

An Overview of the TI-92

A. The Nine *TI-92* Arenas

There are nine screen types available on the *TI-92*. They are: the HOME screen, the Y = Editor, the Window Editor, the Graph screen, Table, the Data/Matrix Editor, the Program Editor, the Geometry screen, and the Text Editor. Each of these nine screens represents a TI-92 application and has a particular function. The HOME screen, for example is where calculations are done and where most mathematical functions are accessed. When you turn the calculator on for the first time, you are in the HOME screen.

We will visit all of these arenas as this overview progresses. As you read through these pages, use *The TI-92 Guidebook* that accompanies your *TI-92* to get a firm understanding of how to use your calculator to its fullest!

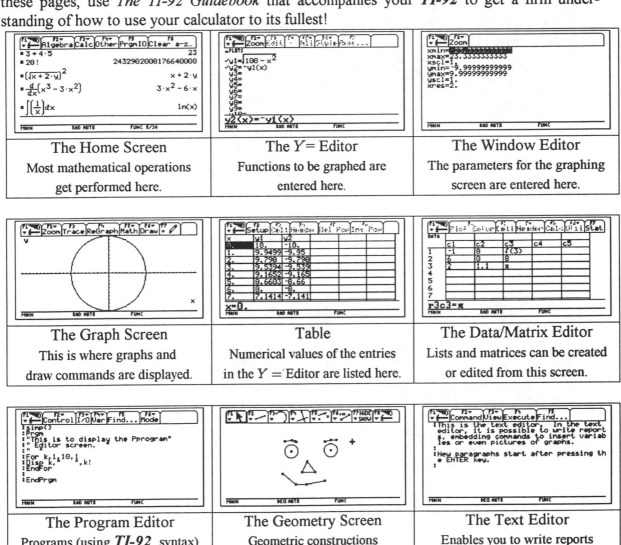

The Home Screen	The Y= Editor	The Window Editor
Most mathematical operations get performed here.	Functions to be graphed are entered here.	The parameters for the graphing screen are entered here.

The Graph Screen	Table	The Data/Matrix Editor
This is where graphs and draw commands are displayed.	Numerical values of the entries in the Y = Editor are listed here.	Lists and matrices can be created or edited from this screen.

The Program Editor	The Geometry Screen	The Text Editor
Programs (using *TI-92* syntax) are written and edited here.	Geometric constructions are created in this screen.	Enables you to write reports or command scripts.

FIGURE A.1.
The Nine *TI-92* Arenas

B. The MODE Key

Let's begin by becoming acquainted with the MODE key -- it is the control center for the appearance of many things in the *TI-92* 's screen displays. There are two pages of information available when you press the MODE key. (Remember, to return to the Home screen at any time, press the green diamond key ◇ followed by the HOME key. We display the options in the two figures below:

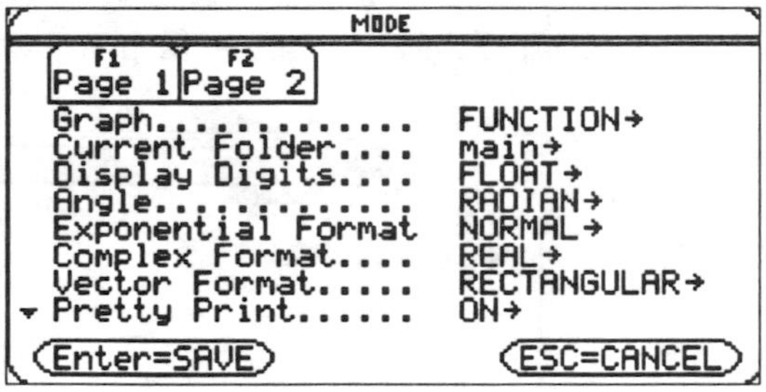

Graph: Select the type of graph from five options: function, function, parametric, polar, sequence, and 3D.

Current Folder: Choose folder name.

Display Digits: Choose the number of digits displayed (from 0 to 12) and fixed or floating point precision.

Angle: Choose radian or degree measure for trig functions.

Exponential Format: Normal, scientific or engineering

Complex Format: Select real or complex numbers. Complex can be in either rectangular or polar formats.

Vector Format: Displays 2 and 3 dimensional vectors in rectangular, cylindrical, or spherical coordinate systems.

Pretty Print: Controls display of fractions and exponents. It can be either ON or OFF.

FIGURE B.1.

Page 1 of the MODE key.

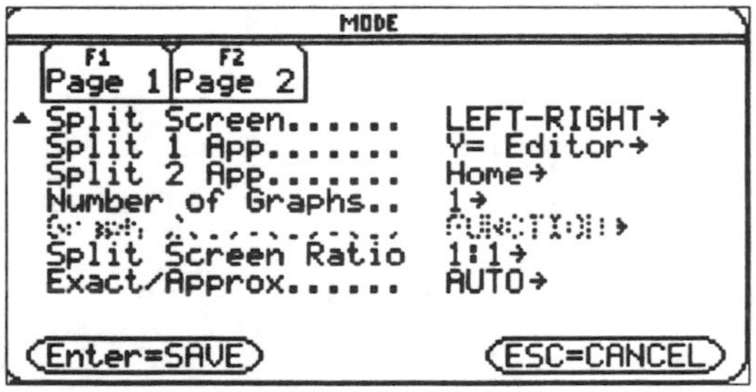

Split Screen: Select from three formats: full screen, top-bottom, or left-right splits.[1]

Split 1 App: Choose from nine possible screens: Home, Y = Editor, Window Editor, Graph, Table, Data/ Matrix Editor, Program Editor, Geometry, Text Editor.

Split 2 App: Choose from the same list of 9 as in the Split 1 App option.

Number of Graphs: Select 1 or 2.

Graph 2: Select type of second graph (if the number of graphs is 2)

Split Screen Ratio: Choose 1:1, 1:2, or 2:1 (if split screen is selected above).

Exact/Approx: Select AUTO, EXACT, or APPROXIMATE for display of fractional and symbolic results.

FIGURE B.2.

Page 2 of the MODE key.

---

[1]Note: If "FULL" screen is selected, several of the options will not be accessible. They will be ghostlike in appearance. These options are "Split 2 App:", "Number of Graphs", "Graph 2", and "Split Screen Ratio".

Let's choose some MODE options. After pressing the MODE key, press [F2] to obtain the second page of settings. To see the options available under "Split Screen" press the cursor key to move to the right.[2] You should see the three options:

```
1:FULL
2:TOP-BOTTOM
3:LEFT-RIGHT
```

To pick the "3:LEFT-RIGHT" option, either press the numeric [3] key or "cursor down" to the entry and press [ENTER]. Next, after pressing the cursor key to move down one row, select Home for the Split 1 Application and (after you "cursor down" to the next row) select the Y = Editor for the Split 2 Application. You should see the options listed in Figure B.3. To accept or save these, press [ENTER] one more time to see the result of your choices in Figure B.4.

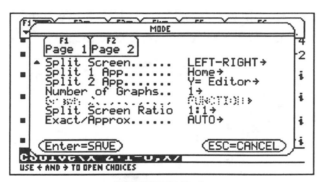

FIGURE B.3.

To select these options, press [ENTER] to see the split screen of Figure B.4. Exiting the MODE screen by pressing [ESC] would not save your selected settings

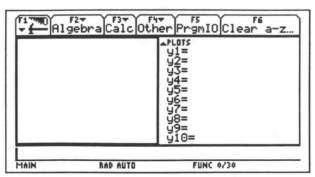

FIGURE B.4.

The Home Screen is on the left and the Y = Editor is on the right.

For an example of how the mode key can affect the appearance of numerical results, let's examine two different digit display possibilities. First, from page one of the MODE options, select FLOAT under the "Display Digits" option; then from page 2 of the settings,

---

[2]When we say "move the cursor to the right" we are, of course, referring to the circular cursor movement key. If you think of this key as a circular clock, "to the right" can be equated to pressing the 3 o'clock position.

select FULL screen.   Now Press [ENTER]. From the HOME screen, enter $2/7 + 3/8$ on the entry line and press [ENTER]. As you can see below, the result is 37/56.

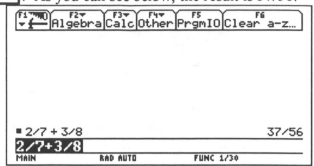

FIGURE B.5.

But if we had wanted the result expressed as a decimal, we would have entered 2/7+3/8 followed by [◇] [ENTER] (for an *approximate* answer). See the result, displayed with 12 decimal places, in Figure B.6.

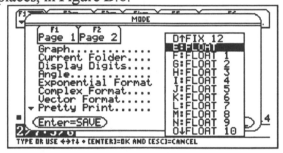
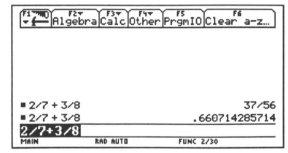

FIGURE B.6.
The approximate result of $2/7 + 3/8$ in FLOAT mode: .660714285714.

Now, from the Display Digits option of [MODE], select and save the FIX 3 option. (You may have to scroll up or down to see it; it is item #4: 4:FIX 3.) Since $2/7 + 3/8$ is still on the entry line, all you have to do is press [◇] [ENTER] when you are back at the HOME screen. The result of 2/7+3/8 is approximated and displayed with three decimal digits in Figure B.7.

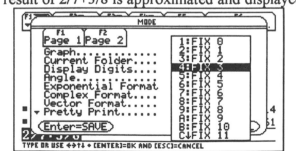
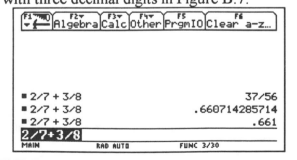

FIGURE B.7.
The result of $2/7 + 3/8$ in FIX 3 mode: .661.

As another example, compare the result of multiplying 0.70145899987 times 100000. Internally, there is no difference in the calculation but the display varies depending on your choice in the mode screen. In Figure B.8, FLOAT 4 is compared with FIX 4:

| ■ .70145899987 · 100000 | 7.015ᴇ4 | ← FLOAT 4 mode |
| ■ .70145899987 · 100000 | 70145.9000 | ← FIX 4 mode |

FIGURE B.8.

Note also the difference in displayed results depending upon the appearance of the data on the entry line. A decimal point in one or more expressions on the entry line causes the result to be displayed with a decimal point. So, for instance, $1/10 + 1/2$ produces output of $3/5$, while $.1 + 1/2$, $1/10 + .5$, and $.1 + .5$ each produce .6 as the displayed result (in FLOAT mode).

C. Feeling at Home on the HOME Screen: Arithmetic Operations

Especially as you begin learning the power of the *TI-92*, you will do much of your work in the HOME screen. Be prepared to see a different type of result than you have experienced with other calculators! Remember the *TI-92* is a symbolic calculator and whenever it can, the *TI-92* will try to return an *exact* result. For work in this section, we will assume that "AUTO" has been selected in the MODE settings under "Exact/Approx", "FLOAT" under "Display Digits" and "RADIAN" under "Angle". Let's examine the difference between exact and approximate displayed results.

For a first example, consider the problem: $2/3 + 4/5$. If we enter that calculation on the entry line and press ENTER the exact result of $22/15$ is displayed. We can even express the result as a mixed number by pressing the F2 Algebra menu and selecting "7: propFrac(".[3] See Figure C.1 below:

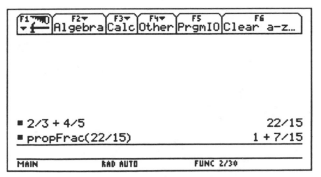

FIGURE C.1.
Notice the exact nature of these results!

---

[3]To recall the last result (22/15), use the cursor keys to move up into the history area. When the entry you wish to recall is highlighted, press ENTER and it will appear on the entry line at the insertion point. Now just press)| then ENTER to view the result.

Keep in mind that if we enter $2/3 + 4/5$ and press ⬦ [ENTER] ($\approx$)the result is approximated and displayed in decimal form.

As a second example of the comparison between exact and approximate results, let's enter $\sqrt{3+4}$. The keystrokes for this are: [2nd] [√] 3 + 4 [)] [ENTER] . We see in Figure C.2 that the *TI-92* returns $\sqrt{7}$.

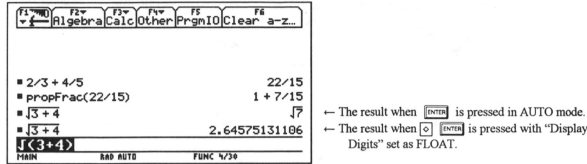

← The result when [ENTER] is pressed in AUTO mode.
← The result when ⬦ [ENTER] is pressed with "Display Digits" set as FLOAT.

FIGURE C.2.
Exact and approximate results for $\sqrt{3+4}$.

Let's evaluate some trigonometric functions. To find the $\cos\left(\frac{\pi}{2}\right)$, first press the **cos** key. That will display "**cos(**" on the entry line. Then press the following keystrokes:

[2nd] [π] [÷] 2 [)] [ENTER]

The *TI-92* displays the result, 0. In Figure C.3, that result as well as the value of sin(.1) and the value of $\tan\left(\frac{3\pi}{4}\right)$ are displayed.

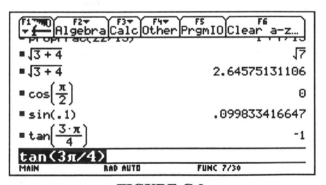

FIGURE C.3.
Three trigonometric calculations.

D. Algebra on the HOME Screen

Symbolic expressions can be entered into the *TI-92* -- not just numerical ones. When a symbolic expression is entered, the *TI-92* attempts to simplify the expression. For example, try entering $\sqrt{x^6}$. The keystrokes for this are [2nd] [√] [x] [^] 6 [)] [ENTER] . Notice that the *TI-92* returns the correct simplification $|x^3|$ in Figure D.1.

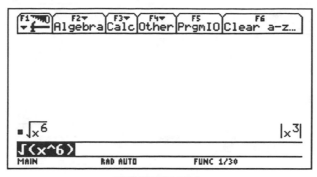

FIGURE D.1.

By pressing the ⬚F2⬚ key from the HOME screen, we can access a menu of eleven algebra functions. These eleven options from the "⬚F2⬚ **Algebra**" menu appear in Figure D.2.

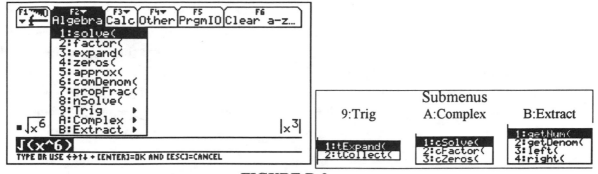

FIGURE D.2.

The eleven algebra options. The " ▷ " indicates that there are submenus for 9:Trig, A:Complex, and B:Extract.

For a first example using the algebra options, let's have the *TI-92* solve the equation $x^2 - 5x + 3 = 0$ for x. From the HOME screen, press ⬚F2⬚ for the algebra options and then select "1:solve(". Immediately, the six characters, solve(appear on the entry line. Now enter the equation followed by ⬚,⬚ ⬚x⬚ ⬚)⬚ as you see on the entry line in Figure D.3.

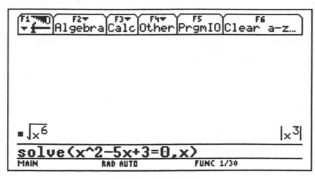

FIGURE D.3.

Now, press ⬚ENTER⬚ to have the *TI-92* display the two solutions in exact form as you can see in Figure D.4

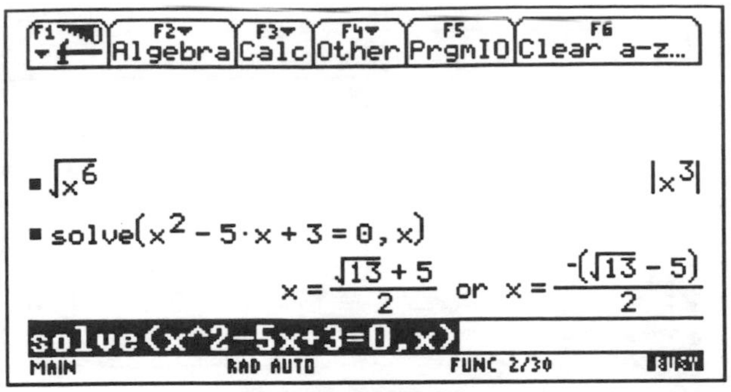

FIGURE D.4.

The two solutions to $x^2 - 5x + 3 = 0$ are $\dfrac{5 \pm \sqrt{13}}{2}$. Notice how the *TI-92* displays them *exactly*.

We point out again that the screens shown here have come directly from a *TI-92* that was in "AUTO" mode. If the calculator was in "Approximate" mode or if we pressed ◇ ENTER , we would have the numbers 4.30277563773 and 0.697224362268 displayed. In fact, a third way to have the approximate solutions to this equation displayed is to use the F2 **Algebra** option "5:approx(" with the solve command as Figure D.5 shows. To try this you can select the "approx(" option then the "solve(" option and then reenter the original equation and press ENTER . Another way to do this while solve $(x^2 - 5x + 3 = 0, x)$ is still on the entry line is to use the cursor key to place the insertion point all the way at the beginning of the line. Then select the "approx(" option, move the insertion point all the way to the end of the line (2nd right cursor) and then press) and ENTER .

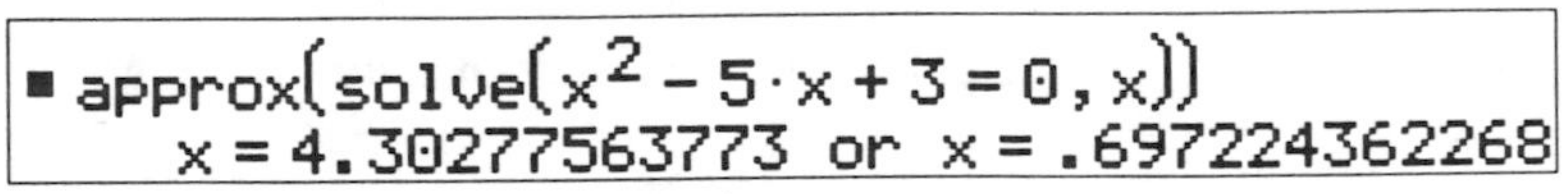

FIGURE D.5.

We will not access all eleven algebra menu options here but we will examine a few more. There are several variations of factoring on the *TI-92*. First, we decide whether we wish to factor over the real numbers or the complex numbers. The "2:factor(" selection on the algebra menu is used for factoring over the real numbers. (Expressions which factor over the complex numbers require the cFactor submenu option after selecting "A:Complex".)

We can factor numbers or symbolic expressions. The *TI-92* returns the prime factorization of the number n when we enter "factor(n)". The first line in Figure D.6 is an example of factoring a number. We entered the command factor(20!) and the *TI-92* returned the prime factorization of 20!.[4]

The remaining four lines on the left screen of Figure D.6 show other variations of the factor (and cFactor) commands. Notice that the general syntax is "factor($expr[, var]$)" which means that entering a variable after the expression is optional. In general, it is advisable to put the appropriate variable name in. You may want to examine the difference between the returned expressions for "factor($x^2 - 5$)" and "factor($x^2 - 5, x$)".

---

[4]The factorial operator, !, is most easily accessed by pressing the 2nd W key.

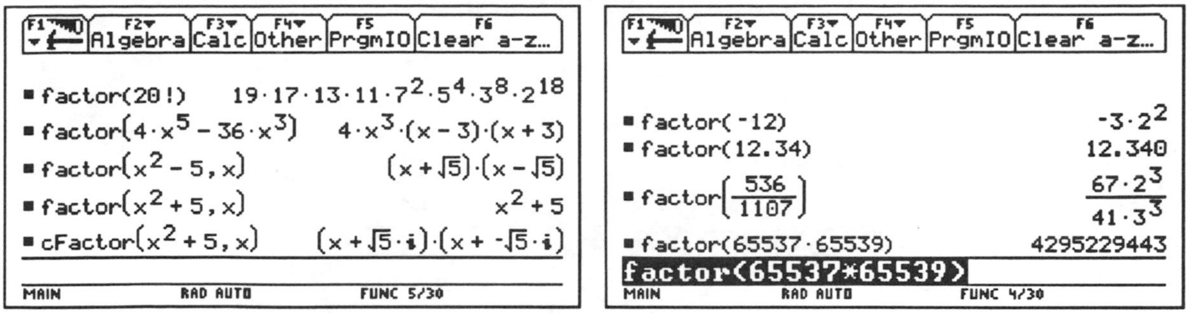

FIGURE D.6.
Factoring variations.[5]

Consider the expression $b^2x^6 - 2b^2x^4 + b^2x^4$. If factored as though x is the variable (and b is a constant), one result occurs and is displayed on the second line of Figure D.7. If we consider b as the variable (and treat x as a constant) the result is different (bottom line of Figure D.7).

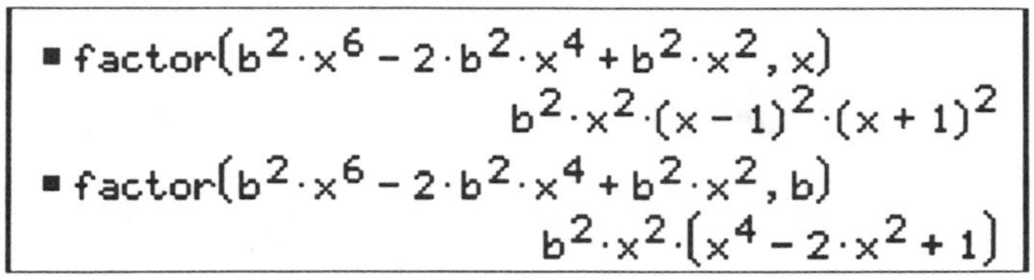

FIGURE D.7.

We conclude this section by demonstrating some of the real power of the *TI-92*. The "zeros(" command is used to return a list of possible real zeros of an expression. The "cZeros(" command does the same for complex zeros. Consider the expression $e^{-x^2} - .5$. In AUTO and EXACT modes, the *TI-92* will try to return zeros in an exact (rational) form. But as you can see below, it only displays the result in decimal form. When the expression is written as $e^{-x^2} - \frac{1}{2}$, however, the output now is in EXACT (rational) form.[6]

The "cZeros(" command is located in a submenu under the "A:Complex" option of the Algebra menu. Select it and [ENTER] and then complete the entry line with x^3 + 1,x) [ENTER] to find its three zeros in exact form.

---

[5]We include factor(65537*65539) to demonstrate that any finite calculation tool has its limitations. The *TI-92* cannot factor the product of the twin primes 65537 and 65539 because the greatest prime factor it can find is 65521.

[6] The exact variation is shown here:

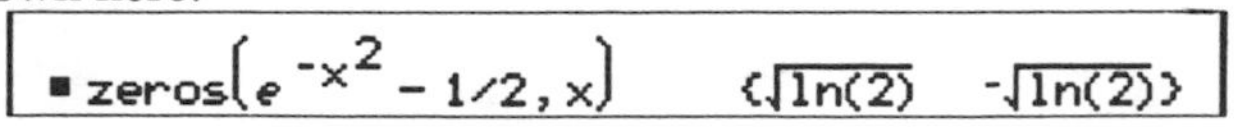

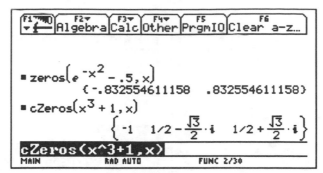

FIGURE D.8.

The two approximate zeros of $e^{-x^2} - .5$ and the three exact zeros of $x^3 + 1$.

E. Some Editing Tips

There are a few things that can save you time when entering and changing entries on the HOME screen. Below, we list a few that have helped us.

1. On a split screen, the 2nd APPS key toggles the active screen from one side to the other.

2. The backspace key, ←, deletes the character to the *left* of the cursor. BUT, ◇ ← is sometimes more convenient -- it deletes the character to the *right* of the cursor.

3. To clear the HOME screen and all previous entries, press F1 and then choose "8:Clear Home". Pressing the CLEAR key will erase the line you are on when you press it. (If the cursor is not at the end of the line, you must press CLEAR twice to clear the edit line.)

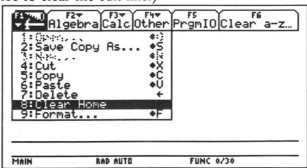

FIGURE E.1.

The F1 menu.

3. Select an item in a pull-down menu by merely typing its associated number (or letter). For example, once the F1 key is pressed and its menu drops down, for the Clear Home function simply press the numeric 8 key. (See Figure E.1.)

4. Instead of searching various menus for a particular function or command, it may be easier to use the 2nd CATALOG listing of all the *TI-92*'s built-in functions and instructions. Once the catalog is selected press the first letter of the function you want and the *TI-92* displays the (alphabetical) catalog listing starting with that letter. If you need a symbol, scroll back from the "A" section or down from the "Z" section to see the symbols in the catalog.

5. When you are editing on the EDIT line, pressing the 2nd key before moving left (or right), will put the cursor at the extreme left (or right) of the expression being edited. Similarly, while in the CATALOG (or in any other long menu), pressing the 2nd key before moving the cursor up or cursor down will move a *page* at a time.

6. If an item in a pull-down menu has a ▷ to its right, then that item has a sub-menu.

7. Get used to copying and pasting. To copy a highlighted expression,[7] simply press ◊ C . Then move the cursor to where you want to place this expression and press ◊ V to paste it. This is particularly useful if an expression on the HOME screen is to be copied into the $Y =$ editor, for example.

8. To recall a previous entry it is often easiest to scroll up through the history until it is highlighted and then press ENTER to copy it to the edit line. An alternative to this is to use the 2nd ENTRY key. Similarly, 2nd ANS will recall the last answer returned but you may find it more convenient to scroll the history of entries.

F. Graphing Functions and Displaying Tables

It is quite easy to get started graphing functions of a single independent variable on the *TI-92*. We begin by observing three important graphing-related keys: ◊ Y= , ◊ WINDOW and ◊ GRAPH. The first function we will graph is $f(x) = (x + 2)(x - 1)^2$. To do this, press ◊ Y= and use the cursor key to select an "available" function, say $y1$. Now press ENTER and an insertion bar appears next to $y1(x) =$ on the entry line. Type in the expression $(x + 2)(x - 1)$^2 followed by ENTER. So, we have entered our function as $y1(x)$ (not as $f(x)$).

Now set the WINDOW settings to their default settings by pressing ◊ WINDOW, then F2 and then choose "6:ZoomStd".

---

[7]To highlight an expression, hold the shift key down while moving the cursor over it.

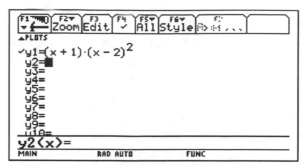

FIGURE F.1.

$y1(x)$ is entered as 〔(〕〔X〕〔+〕〔1〕〔)〕〔(〕〔X〕〔–〕〔2〕〔)〕〔^〕〔2〕〔ENTER〕 .

Using the default window values, by pressing ◇ GRAPH , we see a **TI-92** rendition of the graph. (To view the window values in effect, simply press ◇ WINDOW .) Those window values as well as the graph are displayed in Figure F.2 below.

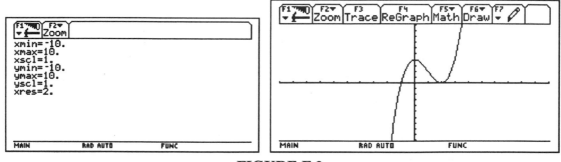

FIGURE F.2.

Note the distance between tick marks on the x and y axes is not equal when using the default (also called **ZoomStd**) window values, even though each tick is one unit on both axes.

Before we graph additional functions or examine different window settings, it should be noted that there are several graph formats that can be used to view graphs. These format variations are found in the graph screen under the F1 key. After pressing F1 press 9 to access the "9:Format" option -- you will see the screen in Figure F.3. You can choose among three choices for Coordinates (RECTangular, POLAR or OFF), two choices for Graph Order (SEQuential or SIMULTaneous), two choices for Grid (ON or OFF), two choices for Axes (OFF or ON), two choices for Leading Cursor (OFF or ON), and two choices for axis Labels (OFF or ON).

We select "Grid" by moving the cursor down to the third line and then right-cursor to see the menu of : "1:OFF 2:ON". Cursor-down or press the 2 key to select "2:ON". Then press ENTER twice to return to the GRAPH screen.

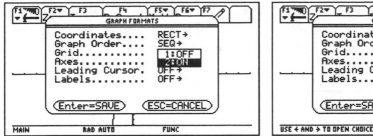 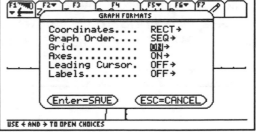

FIGURE F.3.

Select Grid "ON" and the graph of Figure F.2. is redrawn with grids in Figure F.4.

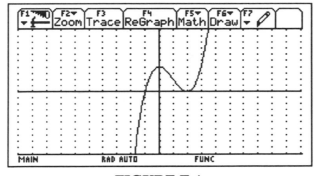

FIGURE F.4.

Graphs can also be captioned. Press the F7 key while viewing the GRAPH screen to see the menu of drawing options. Option "7:Text" can be used to write text on the screen. After selecting "7:Text", move the cursor to the location of the first character to be displayed and then just start typing the text. One use of captioning is shown in Figure F.5.

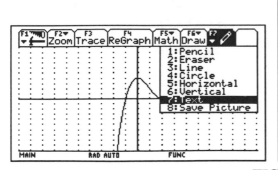 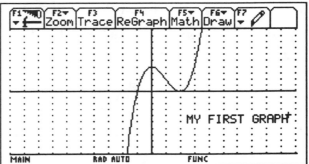

FIGURE F.5.

Captioning a graph with text.

To clear any drawn items (like the text above) from the GRAPH screen, press F6 then select "1:ClrDraw" and then press ENTER. Or, after pressing F6, press 1 to select and execute the ClrDraw feature.

The *TI-92* can also show a table of values. We will continue to use the function $f(x) = (x + 2)(x - 1)^2$ as $y1(x)$. But we want a second function, $y2(x) = 2x - 1$, and we enter that in the $Y =$ Editor as shown in Figure F.6 below.

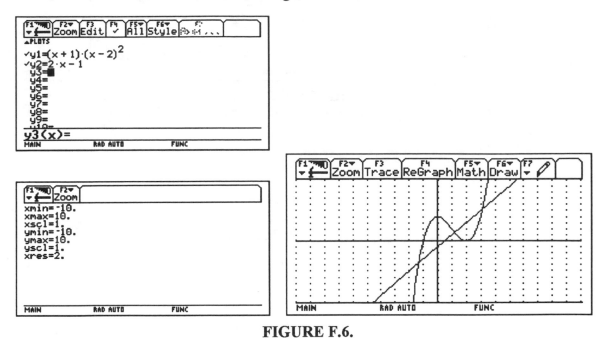

FIGURE F.6.
Two functions are graphed on the same graph screen.

We would now like to display a table of numerical values that represent points on these graphs. Begin by setting up the table parameters by pressing the $\diamond$ [TblSet] key. Two important values have to be put into the table: a starting x value (labeled tblStart) on the TblSet screen, and an x-increment value (labeled as Δtbl). For this example, let's start the table at $x = -5$ and increment the x values by 0.5. We enter these values as in Figure F.7 and press [ENTER] to save them and exit the TblSet screen. Now, press $\diamond$ [TABLE] to view the table displayed in Figure F.8.

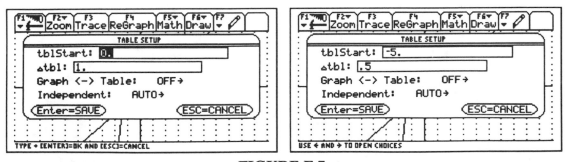

FIGURE F.7.
The TblSet Screen: the default values (left) are changed to get the table in Figure F.8.

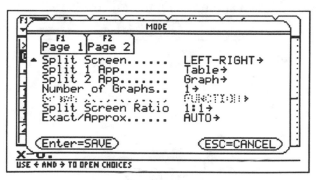

FIGURE F.8.

These table values can be scrolled upward or downward.

An interesting view occurs when the table and graph appear adjacent to each other. To do this we must split the screen using the MODE key. Notice in Figure F.9 that we chose to put the table on the left ("Split 1 App") and the graph on the right ("Split 2 App").

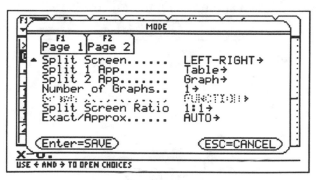

FIGURE F.9.

Page two of the MODE options.

To exhibit values in the table that correspond with points graphed, choose "ON" for the Graph ↔ Table option after pressing ◇ TblSet. This will adjust the tblStart and Δtbl values to correspond with the *x*-values in the graph window.

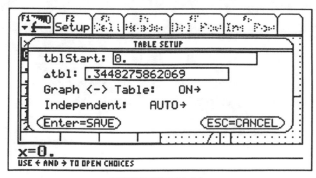

FIGURE F.10.

With the Graph ↔ Table option "ON", values for *x* will correspond to graph screen *x*-values.

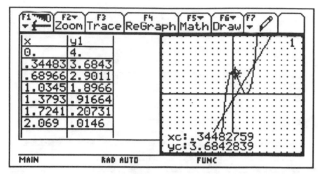

FIGURE F.11.

The split screen of Table and Graph. We also selected $\boxed{F3}$ Trace to show that coordinates match table values.

By now you should feel comfortable enough with the **TI-92**'s $\boxed{\text{MODE}}$ settings so that you can "unsplit" the screen[8] and enter a new function into the $Y =$ Editor. This new function, $y3(x) = \frac{x^2-4}{x+2}$ is the only one we want to graph. So, we can deselect $y1(x)$ and $y2(x)$ by placing the cursor on each of them and pressing the $\boxed{F4}$ key.

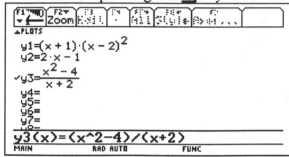

FIGURE F.12.
Only $y3$ is checked so only $y3$ will be graphed.

Now, the observant mathematics student may realize that this function has a restricted domain. That is, $x \neq -2$. But if this graph is drawn with the standard window values (as in Figure F.2) it will appear to be a straight line. There cannot be a point on the graph, however, at $x = -2$. To help "force" the **TI-92** to graph at precisely $x = -2$ (thereby showing the "hole"), we need to know something about pixels. Pixels (or "picture elements") are the spots that are lit on the screen. In an unsplit graph screen, there are 239 pixels from left to right and 103 from top to bottom.[9]

---

[8]Do this by selecting "1:FULL" from the Split Screen submenu of the $\boxed{\text{MODE}}$ key.
[9]It is necessary to know pixel ranges for each possible graph screen. Although they are listed in Chapter 5 of the *TI-92 Guidebook*, we repeat them here for easy reference:

Full Screen	0-238 x	0-102 y
Top/Bottom Screen		
Split 1:1	0-234 x	0-46 y
Split 1:2	0-234 x	0-26 y
Split 2:1	0-234 x	0-68 y
Left/Right Split Screen		
Split 1:1	0-116 x	0-98 y
Split 1:2	0- 76 x	0-98 y
Split 2:1	0-156 x	0-98 y

If we choose a range of x-values that evenly divides into 238, pixels will fall on "nice" numbers. For example, if xmin $= -5$ and xmax $= 6.9$, that range of 11.9 divides into 238 evenly and the result of that division is 20. In other words, there will be twenty pixels for each unit in the x-direction. Equivalently, each accessible x-value is $1/20 = 0.05$ units.

To modify the default graphing window settings, press $\boxed{\diamond}$ $\boxed{\text{WINDOW}}$ and, using the cursor key, highlight the values you wish to change and enter new values. Often, you may have to do this several times before an appropriate viewing window is obtained. You can see our choices in Figure F.13. Notice also in Figure F.13 that we have displayed part of a table of values for x and $y3$ (remember, $y3 = \frac{x^2-4}{x+2}$). To produce these values, press $\boxed{\diamond}$ $\boxed{\text{TblSet}}$ and enter a starting x value. Since our viewing window starts at $x = -5$, we use this value for tblStart. For the Δtbl value be sure to choose a value which will ensure that -2 appears in the table of x values. Suitable choices include .05, .10, 1, and .2. We chose .1. After saving and exiting the TblSet screen, view the table and use the cursor key to scroll down through the table pausing at the x value of -2. Notice that the dependent variable $y3$ is undefined there.

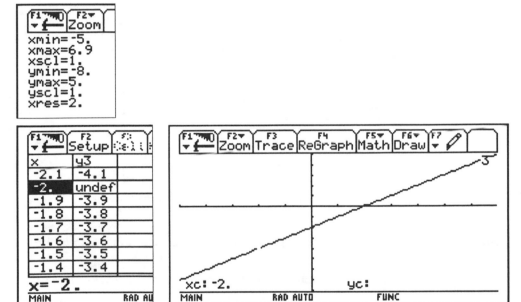

FIGURE F.13.[10]
Both the table and the graph's "gap" indicate there is no function value at $x = -2$.

Remember, it is often handy to have the table and graph together, side-by-side, on the same screen. To accomplish this, use the $\boxed{\text{MODE}}$ key to split "LEFT-RIGHT". For another example of a function with a point "missing", we enter $\dfrac{x^3 - 1}{x - 1}$ for $y1(x)$ in the $Y =$ Editor. To clear all existing functions from the $Y =$ editor, select "8:Clear Functions" from the $\boxed{\text{F1}}$ menu of $Y =$. To clear the HOME screen, select "8:Clear Home" from the $\boxed{\text{F1}}$ menu of HOME. Also, keep in mind that to switch from one screen to the other while in split screen

---

[10]We have turned off the grid by selecting that option after pressing the $\boxed{\text{F1}}$ key and choosing "9:Format".

mode, press $\boxed{2^{nd}}$ $\boxed{\text{APPS}}$ -- that is, the second function of the $\boxed{\text{APPS}}$ key. Finally, we mention that since we are in a split screen and we want to see what occurs at $x = 1$, we choose window values that will force $x = 1$ to coincide with a pixel .

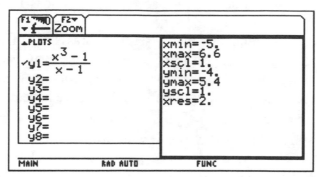

FIGURE F.14.

$\dfrac{x^3 - 1}{x - 1}$ is equivalent to $x^2 + x + 1$ except at $x = 1$. Notice the range of x-values, $6.6 - (-5) = 11.6$, corresponds with 116 pixels so that each unit in the x direction is made up of 10 pixels.

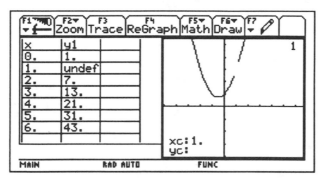

FIGURE F.15.

There is a gap at $x = 1$.
Notice that in the left window, we have displayed TABLE values in which Δtbl $= 1.0$.

We conclude this section by examining how to vary the appearance of a graph. First, unsplit the screen. In the $Y =$ Editor, the $\boxed{\text{F6}}$ Style key allows the user to assign one of eight different styles to a graph.

1:Line	Default (all previous graphs in this section have been "Line")
2:Dot	Points are not connected (recommended for graphs with asymptotes)
3:Square	A solid box is graphed at each point.
4:Thick	A thick line is drawn
5:Animate	A circle follows the path for the curve but nothing is drawn
6:Path	Like animate but the trail of the path followed *is* drawn
7:Above	The curve is drawn and the area above the curve is shaded
8:Below	The curve is drawn and the area below the curve is shaded

FIGURE F.16.

The Style options.

Consider graphing the functions given by $y1 = 4 - x^2$ and $y2 = 2x + 1$. Select the desired style while each function is highlighted in the Y = editor. (We chose "4:Thick" for $y1$ and "2:Dot" for $y2$.) Then choose "6:ZoomStd" from the F2 Zoom menu. Figure F.17 shows this.

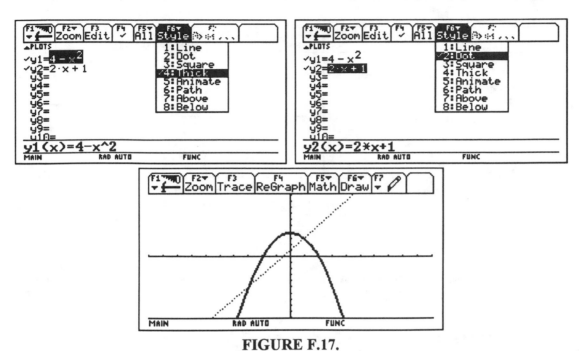

FIGURE F.17.
The styles of "Thick" for $y = 4 - x^2$ and "Dot" for $y = 2x - 1$ are displayed.
(Standard window values were used.)

G. The F2 Zoom Graph Menu

By pressing the F2 key from the *TI-92* Graph screen, you access twelve different zoom variations. Figure G.1 displays them. We will discuss these variations in this section.

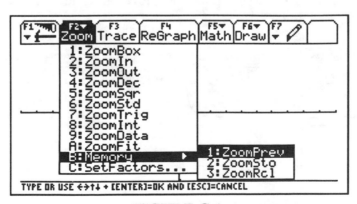

FIGURE G.1.
The twelve options of the F2 Zoom as well as three submenu options under "B:Memory."

G.1. ZoomBox

For this example, we want to graph the function $f(x) = \dfrac{x^2 + 1}{x^2 - 4}$ and examine where a relative maximum value occurs near $x = 0$. An initial plot appears in Figure G.2 below.

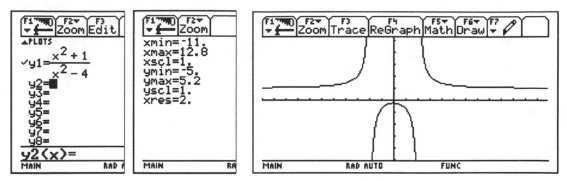

FIGURE G.2.

An initial look at $f(x) = \dfrac{x^2 + 1}{x^2 - 4}$ in Style "line".

Now, we select ZoomBox from the F2 menu and proceed by creating a box around the point on the graph where $x = 0$. It is a three step process:

Step 1: After selecting "1:ZoomBox" from the F2 menu, locate the cursor at the upper left-hand corner of the box that is about to be created. We chose the point $(-2, 0.3)$. Press the ENTER key.

Step 2: Now move the cursor to the location of the lower right-hand corner of the box. We chose the point $(2, -1.5)$. Press ENTER.

Step 3: The *TI-92* redraws the graph using the box as the viewing window. It is always good to observe the new window values after doing a ZoomBox. See Figures G.3 and G.4.

Step 1: Locate the first corner and ENTER   Step 2: Locate the second corner and ENTER   Step 3: The new graph is redrawn

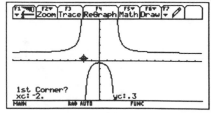 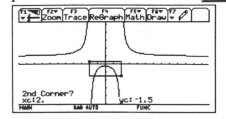 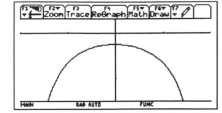

FIGURE G.3.

The three steps of a ZoomBox command.

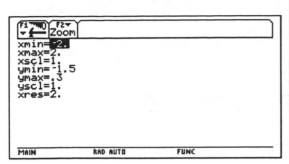

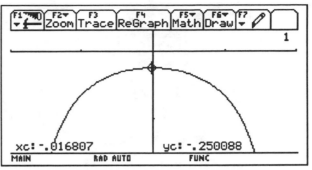

FIGURE G.4.

Notice the new window values after the ZoomBox is completed and [F3] Trace is pressed.

It appears that the maximum y value is near -0.25. Of course, we could calculate the value directly to be $-\frac{1}{4}$ or have the *TI-92* do it directly in the Home screen by evaluating our $y1$ function at $x = 0$ (or by pressing 0 while in Trace mode on the Graph screen):

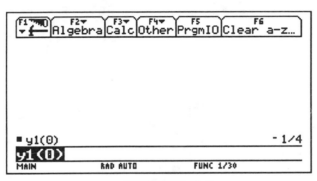

FIGURE G.5.

$y1(0) = -\frac{1}{4}$

G.2. ZoomIn and ZoomOut

The "2:ZoomIn" and "3:ZoomOut" commands in the [F2] Zoom menu are self-explanatory. They are used to zoom in (or out) on a point on a graph. It is important to realize that each of them depends on the values set under the "C:SetFactors…" option of the [F2] Zoom menu. The default is a factor of 4 for both the x and the y direction. We will leave the zoom factors at this default setting for now; they are shown in Figure G.6 below.

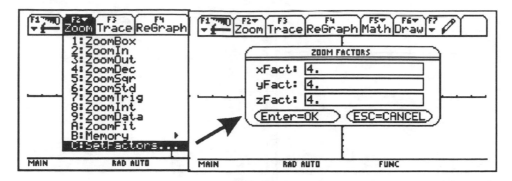

FIGURE G.6.

The default settings for the xFact and yFact values are 4. The zFact setting only pertains to 3D plots.

To see the importance of using the ZoomIn feature, consider the problem of trying to determine whether or not these two functions intersect: $y = 2\cos(x + 1)$ and $y = x^4 + 2x^2 + \frac{3}{2}$. Figure G.7 shows the graphs in standard form:

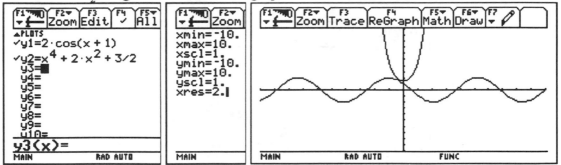

FIGURE G.7.

The graphs of $y = 2\cos(x + 1)$ and $y = x^4 + 2x^2 + \frac{3}{2}$. Do they intersect?
It is difficult to determine using the standard window settings.

Now, with the graph on the screen, we press ▢F2 Zoom and select "2:ZoomIn". Move the cursor to what will be the new center of the graphics window -- we chose the point closest to the apparent intersection, $(-0.252101, 1.568627)$ -- and press ▢ENTER. The new graph gets re-drawn by a zoom-in factor of 4 for each of x and y. Compare Figures G.8 and G.9. Notice that the new scale, 0.25, is one-fourth the old scale (1) because of the zoom factor of 4.

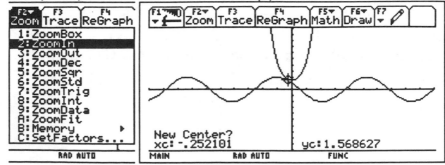

FIGURE G.8.

After selecting "2:ZoomIn", place the cursor to the point where you
want the new center of the graph window to be. Then press ▢ENTER to get the graph in Figure G.9.

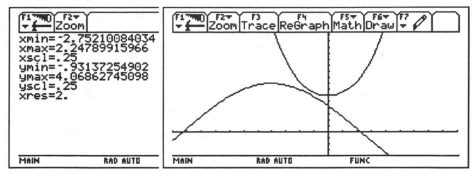

FIGURE G.9.
The result of zooming in shows that the graphs *do not* intersect!

The "3:ZoomOut" option works similarly. It is especially useful when there is no graph visible in the present viewing window.

G.3. ZoomDec

The "4:ZoomDec" (for zoom decimal) option of the zoom menu is used to set the distance between pixels to 0.1 unit. This is particularly useful for tracing curves and having the x-values expressed to only one decimal place. Consider the greatest integer function which the *TI-92* writes as $floor(x)$. Its graph, after a ZoomDec, is displayed in Figure G.10. The reader should try tracing the graph to see the corresponding x values.

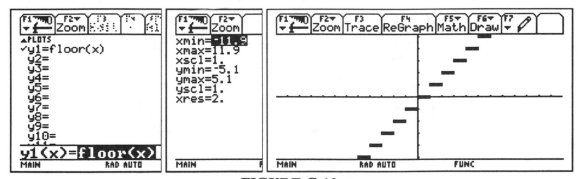

FIGURE G.10.
The greatest integer function (drawn with the square Style).

G.4. ZoomSqr

The equation for a circle of radius r centered at the origin is given by $x^2 + y^2 = r^2$. In the *TI-92* function graphing mode, we need to solve this for y, obtaining the two semicircles, $y = \sqrt{r^2 - x^2}$ and $y = -\sqrt{r^2 - x^2}$. Let's suppose we would like the radius to be $r = 8$. Notice, in Figure G.11, how we could enter these two functions in the $y =$ Editor.

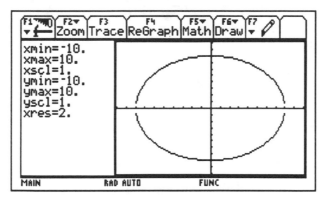

FIGURE G.11.

The use of the "with" parameter $\left(\boxed{2^{nd}} \ \boxed{K} \right)$ temporarily assigns the value of 8 to r.[11]

The graph, using the standard window values and a 1:2 split screen, is shown in Figure G.12. Unfortunately, there is some distortion in the circle due to the scale.

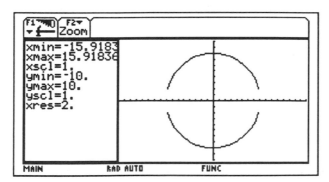

FIGURE G.12.

With these window values, the circle doesn't appear very circular! (The Style is "line" for both halves.)

By selecting "5:ZoomSqr" from the zoom menu, the *TI-92* will adjust the scale so that circles appear circular. Consider the result of the ZoomSqr in Figure G.13.

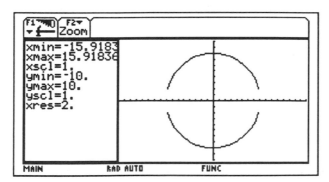

FIGURE G.13.

The circle is no longer distorted. The gap between semi-circles is a result of the way in which the *TI-92* draws graphs internally and cannot be helped.

---

[11]The "with" function is only one of many special characters available directly from the TI-92 keyboard. To see the available special symbols and characters, press $\boxed{\diamond}$ $\boxed{K}$; press $\boxed{\text{ESC}}$ or $\boxed{\text{ENTER}}$ to clear.

G.5. ZoomStd

We have already referred to the "6:ZoomStd" option several times. When this option is invoked, the *TI-92* always (regardless of the way the screen may be split) returns these WINDOW values:

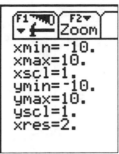

FIGURE G.14.
The window values after ZoomStd is selected.

G.6. ZoomTrig

The "7:ZoomTrig" option sets $xscl = \frac{\pi}{2} \approx 1.5708$, $ymin = -4$, $ymax = 4$, and $yscl = 0.5$. The values of xmin and xmax will depend on whether or not the screen is split and, if it is, what the split ratio is. In any case, the horizontal distance between pixels, Δx, is set to $\frac{\pi}{24}$. Figure G.15 shows the graph of $y = 2\sin(3x - \frac{\pi}{4})$.[12]

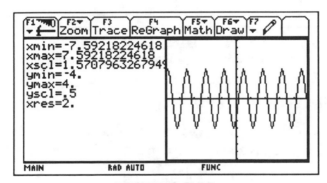

FIGURE G.15.
$y = 2\sin(3x - \frac{\pi}{4})$ after ZoomTrig was selected in a 1:1 LEFT-RIGHT split screen.

G.7. ZoomInt

The "8:ZoomInt" option is similar to ZoomDec but instead of the distance between pixels being 0.1, it is set at 1.0 (for both x and y) after a ZoomInt and after you select a new center. Also, xscl and yscl are each set to 10.

---

[12]To duplicate this screen, be sure to define exactly one function in the Y = editor and then split the screen in the MODE settings. Set Split 1 App to the Window Editor and Split 2 App to Graph. Then, in the left window, select F2 7:ZoomTrig and observe the graph in the right window. Use 2nd APPS to move from one split screen application to the other.

G.8. ZoomFit

The "9:ZoomFit" option is useful for viewing the graphs of functions with either very large or very small ranges. The *TI-92* will retain the current xmin and xmax settings but will change ymin and ymax to accommodate the function's largest and smallest values for x values between xmin and xmax. Figures G.16 and G.17 display the effect of a ZoomFit command for two different functions.

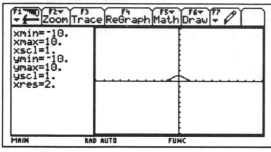 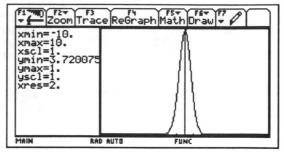

FIGURE G.16.

$y = e^{-x^2}$ after ZoomStd (left) and then after ZoomFit (right) in a 1:2 split screen.

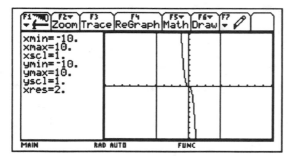 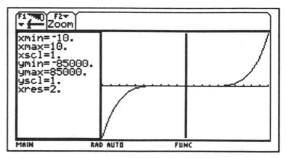

FIGURE G.17.

$y = x^5 - 15x^3$ after ZoomStd (left) and then after ZoomFit (right).[13]

H. The [F5] Math Graph Menu

There are twelve options in the [F5] Math menu that is accessed from the graph screen; they are useful in analyzing graphs and are listed in Figure H.1 below.

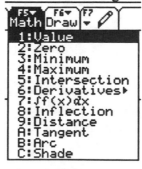

FIGURE H.1.

The twelve options in the [F5] Math menu of the graphics screen.

---

[13]Beware! The ZoomFit command can lose a lot of important detail. Can you explain the appearance of the ZoomFit graph on the right? Try ymin $= -400$ and ymax $= 400$ to see the lost detail.

H.1. Value

The "1:Value" option can be used for evaluating any selected function value for a given x value. For example, if $f(x) = \frac{x}{9-x^2}$ is graphed, we can find the value of $f(1.5)$ by selecting "1:Value" from the math menu and then entering 1.5.[14] The **TI-92** goes into its trace mode and displays the coordinates (1.5, .222222). See Figure H.2 below.[15]

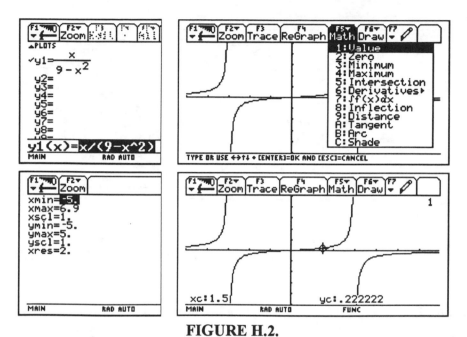

FIGURE H.2.

The value of $y = \frac{x}{9-x^2}$ when $x = 1.5$ is $y \approx 0.222222$.

By pressing $\boxed{\text{MODE}}$ or $\boxed{\text{ENTER}}$ or $\boxed{\text{ESC}}$ the xc: and yc: remnants will be removed from the graphing screen.

H.2. Zero, Maximum, and Minimum

Three of the math options available from the graph screen (2:Zero, 3:Maximum, and 4:Minimum) work similarly. We will demonstrate the Minimum option using the function $f(x) = 2x^6 - 3x^5 - 3x^4 + 2x^3$. An initial plot appears in Figure H.3 below.

---

[14]Keep in mind that you must enter a value between xmin and xmax or you will get a domain error message.

[15]There are two other ways to evaluate our $y1$ function at $x = 1.5$. From the graphics screen, press $\boxed{\text{F3}}$ Trace and then enter the value 1.5 for x (displayed as "xc:" on the screen). Then the y coordinate, yc:, is calculated and displayed. Alternatively, from the Home screen, simply evaluate $y1(1.5)$ as displayed below:

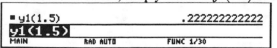

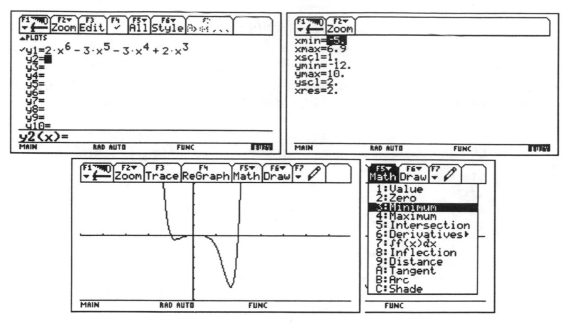

FIGURE H.3.

The graph of $f(x) = 2x^6 - 3x^5 - 3x^4 + 2x^3$. After selecting [F5] Math and then "3:Minimum", you will be prompted for lower and upper bounds on the minimum value of x. (See Figure H.4.)

The **TI-92** now requests that you enter a lower bound. There are two ways to do that: by moving the cursor to an appropriate point (like tracing) or simply by entering an x value. We usually choose the latter method. In this case, we chose the number -5 and then pressed [ENTER]. Similarly, for an upper bound, we entered an x value of 5. See Figures H.4 and H.5.

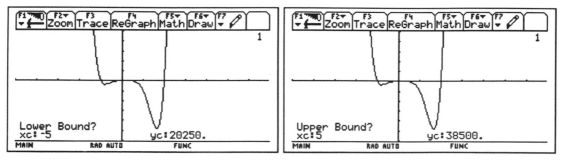

FIGURE H.4.

We enter the lower bound (for x) as -5 and the upper bound as 5.

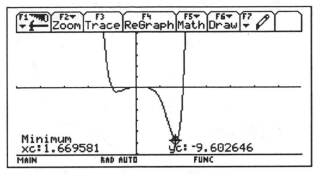

FIGURE H.5.

The minimum value of $f(x) = 2x^6 - 3x^5 - 3x^4 + 2x^3$ is about -9.6 and occurs when $x \approx 1.67$.

H.3. Intersection

The *TI-92* will find the approximate point of intersection of two curves when "5:Intersection" is chosen from the MATH menu on the graphics screen. In Figure H.6, we have entered and graphed the two functions $y = x^{2/3}$ and $y = 4 - x^2$. We want to find the point of intersection that appears to be in the second quadrant.

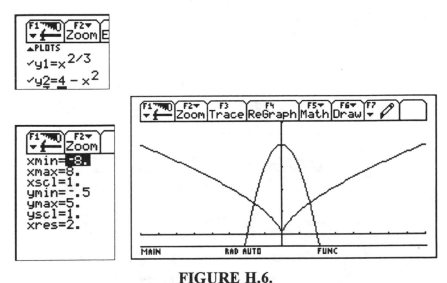

FIGURE H.6.

The graphs of $y = x^{2/3}$ and $y = 4 - x^2$.

After choosing "5:Intersection" from the MATH menu, the *TI-92* prompts you for the first curve, second curve, a lower bound and an upper bound for the point of intersection. After making each selection, press the ENTER key. These four steps are captured in the graphics screens in Figure H.7. The approximate solution to the equation $x^{2/3} = 4 - x^2$ is found to be $x \approx -1.619$ as shown in Figure H.8.

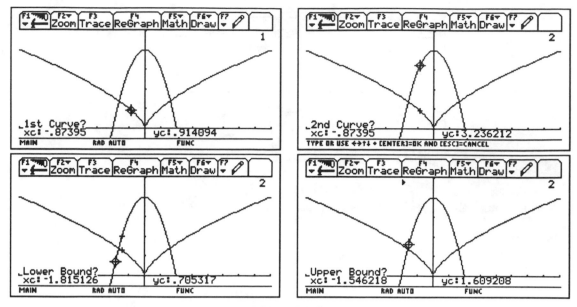

FIGURE H.7.

Above left: press ⌅ENTER⌆ with the cursor anywhere on $y1(x)$. Above right: press ⌅ENTER⌆ with the cursor anywhere on $y2(x)$. Lower left: position the cursor to the left of the apparent intersection and press ⌅ENTER⌆. Lower right: position the cursor to the right of the apparent intersection and then press ⌅ENTER⌆ to get the result in Figure H.8.

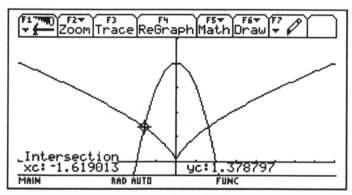

FIGURE H.8.

The graphs of $y = x^{2/3}$ and $y = 4 - x^2$ meet at about the point $(-1.619013, 1.378797)$.

H.4. Derivatives and Tangent

From the graph screen, the *TI-92* will calculate the first derivative of the selected function evaluated at a point. For example suppose $f(x) = 3x^2 - x^3$ is entered as $y1(x)$ in the $Y =$ editor. To find $f'(1)$, select "6:Derivatives" from the Math menu of the graph screen. Then, when prompted for the point, either move the cursor to the desired point or enter the abscissa (x-coordinate) of the point directly from the keyboard. It is generally preferable to enter the value from the keyboard so that you are not limited to x values that fall exactly on pixels. Figure H.9 shows that $f'(1) = 3$.

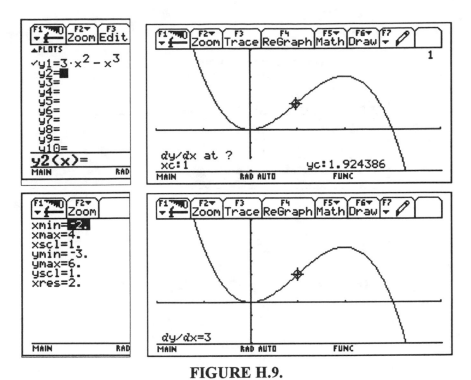

FIGURE H.9.

Finding the value of $\dfrac{dy}{dx}$ from within the graph screen.

A related option on the MATH menu is "A:Tangent." This will draw a tangent line at the point on a selected curve and display the equation of that line. Figure H.10 shows that the equation of the line drawn tangent to $f(x) = 3x^2 - x^3$ at $(1, 2)$ is $y = 3x - 1$.

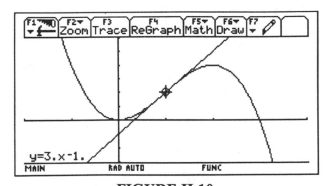

FIGURE H.10.

The equation of the tangent line drawn to $(1, 2)$ is $y = 3x - 1$.

H.5. $\int f(x)\,dx$

Suppose we wish to approximate the area under the curve $f(x) = e^{-x^2}$ between $x = 0$ and $x = 1$. To make this approximation, we enter $f(x)$ as $y1(x)$ in the $Y =$ editor and choose an appropriate viewing WINDOW. Then select "7:$\int f(x)\,dx$" from the [F5] Math menu with the graph displayed. The *TI-92* will prompt you for a lower bound (in this case, we enter 0) and an upper bound (here, $x = 1$). The value of $\int_0^1 e^{-x^2}\, dx$ is approximated and the corresponding area is shaded on the graph as you can see in Figure H.11.

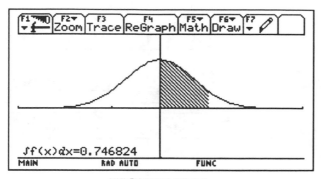

FIGURE H.11.
The approximate value of $\int_0^1 e^{-x^2}\, dx$ is 0.746824.

H.6. Inflection

The *TI-92* is able to locate points of inflection from the math menu of the graph screen. The process is similar to those discussed above. For our example, we will continue with the function $y1(x) = e^{-x^2}$. With the graph displayed, we choose "8:Inflection" from the F5 Math menu. As with other options in this menu, the *TI-92* prompts for a lower bound and an upper bound. Figure H.12 below shows the final screen of this process. By the symmetry of the curve, we can deduce that the two points of inflection occur at $x \approx \pm 0.707107$.[16]

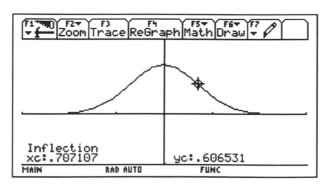

FIGURE H.12.
For $f(x) = e^{-x^2}$, a point of inflection occurs at approximately (0.707107, 0.606531).

This concludes our discussion of a subset of the options available from the F5 Math menu from within the graph screen.

-----

[16]The *TI-92* can find the x-value of the points *exactly* using the "solve(" command on $\frac{d^2(y1(x))}{dx^2}$ from the Algebra menu of the HOME screen. They are $\pm \frac{\sqrt{2}}{2}$.

I. Additional Types of Graphs

In addition to graphing functions of one independent variable, the *TI-92* can be put into modes to do parametric plots, polar plots, plots of sequences and 3D graphs (functions of two independent variables). We will cover these options in this section.

I.1. Graphing Parametric Equations

Select the "2:PARAMETRIC" option of the Graph choices from the $\boxed{\text{MODE}}$ screen. (Be sure to press the $\boxed{\text{ENTER}}$ key twice.)

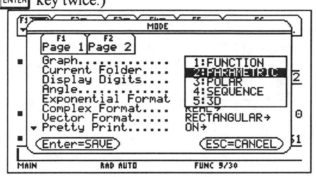

FIGURE I.1.

Choose "2:Parametric" for graph type in the $\boxed{\text{MODE}}$ settings.

Now consider the following problem. If a ball is thrown with an initial velocity of 80 ft/sec at an angle of 40° to the ground, its horizontal component, $x(t)$, and its vertical component, $y(t)$, can be expressed as functions of the parameter t, or time. On the *TI-92*, the Y = key accepts these functions when in parametric graph mode. They are expressed as $xt1$ and $yt1$ for the first parametric pair. In this example, $x(t) = 80\,t\cos(40°)$ and $y(t) = 80\,t\sin(40°) - 16\,t^2$ and they are displayed in Figure I.2 below along with the window values for a graph.[17] When you enter these functions, be sure to use the $\boxed{\times}$ key for explicit multiplication between t and $\cos(40°)$ and between t and $\sin(40°)$.

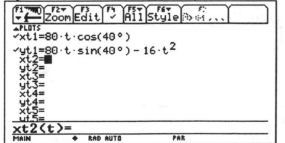 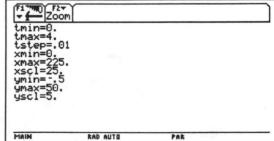

FIGURE I.2.

$x(t) = 80\,t\cos(40°)$ and $y(t) = 80\,t\sin(40°) - 16\,t^2$ are entered in the Y = screen (left) and a suitable collection of window values (including the parameter t) are entered (right).

---

[17]The degree symbol can be entered as $\boxed{2^{\text{nd}}}$ $\boxed{\text{D}}$ directly from the keyboard.

We display the graph represented by these two parametric equations in Figure I.3.

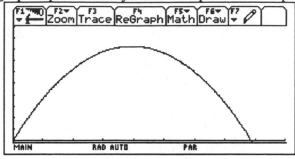

FIGURE I.3.

$x(t) = 80\,t\cos(40°)$ and $y(t) = 80\,t\sin(40°) - 16\,t^2$

To approximate a maximum height for the thrown ball, we can use the trace feature. Notice how the value of the parameter t as well as the x and y coordinates are shown in Figure I.4.

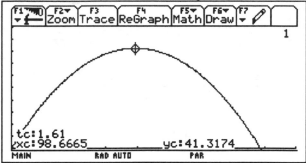

FIGURE I.4.

The maximum height of the ball is about 41.3 feet after about 1.61 seconds.

The graph styles "Animate" and "Path" are often good to choose for parametric equations -- especially for projectile motion effects. With the graph style "Path", we display some of the graphing screens in Figure I.6 below.

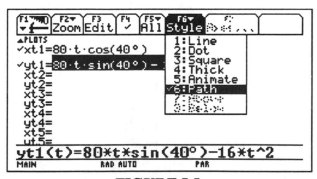

FIGURE I.5.

The style of "6:Path" will be used to simulate the path of the projectile in the next Figure.

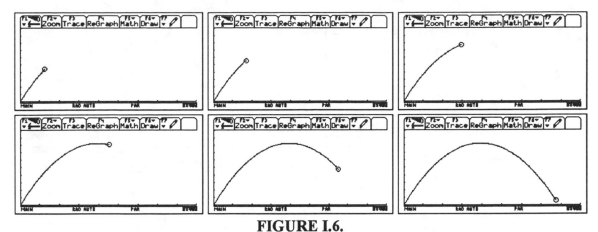

FIGURE I.6.
With the style "6:Path" chosen, we get an idea of the path of the projectile!
Remember, the drawing of a graph can always be halted by pressing the [ON] key or it can be paused by
pressing the [ENTER] key.

I.2. Polar Graphing

The **TI-92** is also capable of graphing a function in polar coordinates where θ is the
independent variable and r is the dependent variable.[18] As an example of such a function, let's
graph $r = 2 \cdot \cos(4\theta)$. First, we set the [MODE] settings to "3:POLAR" as in Figure I.7.

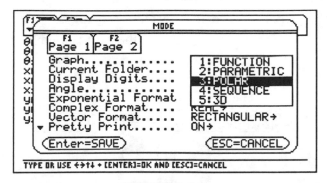

FIGURE I.7.
Choose "3:POLAR" for the Graph option on the first page of [MODE] settings.

Next, we enter the function in the Y = editor. (Notice that the list contains r's not y's
and that the independent variable is now θ.) To obtain the WINDOW settings in Figure I.8,
we issued a ZoomFit command followed by a ZoomSqr command. Then we adjusted θstep to
be 0.05 for a smooth curve. (Compare the smoothness of the graphs in Figures I.8 and I.10.
This smoothness is affected by θstep.)

---

[18]Actually, polar plots can be thought of as a special case of parametric graphing. If $r = f(t)$ represents a
polar-coordinate function, then the parametric representation $x(t) = f(t) \cdot \cos(t)$ and $y(t) = f(t) \cdot \sin(t)$ will
yield the same curve in parametric mode.

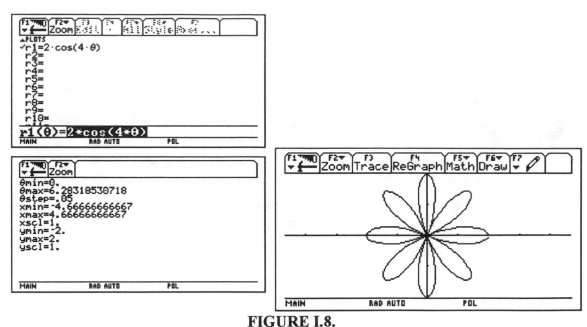

FIGURE I.8.

The polar curve $r = 2 \cdot \cos(4\theta)$, a flower with eight petals.

With the graph displayed, try pressing the TRACE key, [F3]. The default values displayed are the coordinates x, y (displayed as "xc" and "yc") and θ (displayed as θc). Although you can use the arrow keys to trace around the curve, it is also possible to enter a numerical value of θ and after [ENTER] is pressed, the corresponding values of x and y are calculated and displayed. For example, suppose we want to highlight the point on the polar curve for $\theta = 0.25$. We enter it and observe that xc = 1.0470112 and yc = 0.26734586. See Figure I.9.

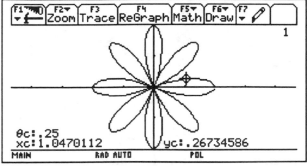

FIGURE I.9.

Choose [F3] TRACE and then enter 0.25. The point (1.0470112, 0.26734586) is displayed in FLOAT mode.[19]

---

[19]It is also possible to have a polar graph displayed with polar coordinates during a trace. To do this, change the format ([F1] 9:Format) to "2:POLAR" under "Coordinates". The r value is displayed as "rc":

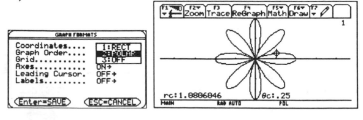

Now, we increase the increment of the graphing radial arm of θ. That is, we increase the value of θstep in Figure I.10 to 0.25. Although the speed of graphing is increased, the smoothness of the curve is sacrificed.

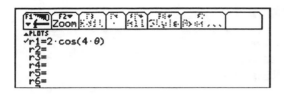

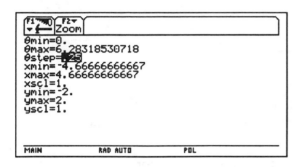

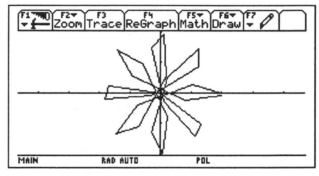

FIGURE I.10.

The polar curve $r = 2 \cdot \cos(4\theta)$ with θstep $= 0.25$.

I.3. Sequence Graphing

A third type of plotting option is "SEQUENCE" and it is selected in the $\boxed{\text{MODE}}$ screen.

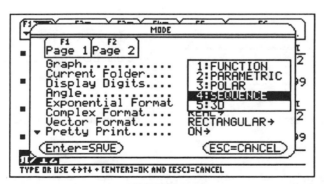

FIGURE I.11.

Choose "4:SEQUENCE" for the Graph option on the first page of $\boxed{\text{MODE}}$ settings

In "sequence" mode, the Y = editor allows us to enter sequences and initial values of sequences. It is possible to enter sequences defined recursively and it is also possible to examine sequences of partial sums.

For a first example, consider the sequence $u1(n) = 1, \frac{1}{4}, \frac{1}{9}, \dots \frac{1}{n^2}$. We enter $u1(n)$ as $1 \div n^2$ in Figure I.12 and also enter 1 as the *initial* value of the sequence (denoted on the

$Y =$ editor as $ui1$). To also examine the sequence of partial sums of the $u1(n)$ sequence, $\sum_{k=1}^{n} \frac{1}{k^2}$ we enter that sequence as $u2(n)$. See Figure I.12.

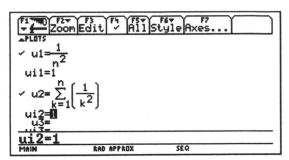 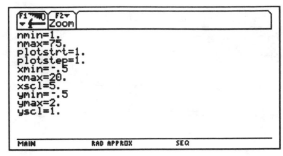

FIGURE I.12.

On the left, the sequence $u1$ is $1/n^2$ and its partial sums, $\sum_{k=1}^{n} \frac{1}{k^2}$ is $u2$. The WINDOW parameters we have

chosen, appear on the right. Notice nmax is 75 (there will be 75 terms calculated and plotted).

We have chosen the "square" style for $u1$ and "thick line" style for $u2$.[20] Each of the initial values is 1 and the graph is drawn in Figure I.13.

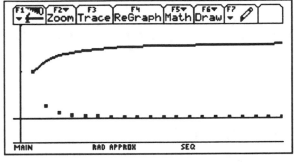

$u2(n)$ is graphed using the "thick line" style.

← The value of this sequence of partial sums approaches 1.6.

← $u1(n)$ is graphed using the "square" style. Notice that the values of the $u1$ sequence approach 0 as n gets large.

FIGURE I.13.

The graph of the sequences listed in Figure I.12. The horizontal axis is n and the vertical axis is u.

If we trace to the last value of n (nmax $= 75$)[21], we can see that $u2(75) \approx 1.631689$. We observe this in Figure I.14.

---

[20]Recall that the styles for each function in the list can be set by using the [F6] key while in the $Y =$ editor.

[21]Just press 75 [ENTER] after pressing [F3] Trace.

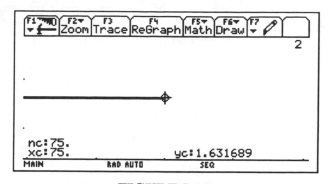

FIGURE I.14.

$$\sum_{k=1}^{75} \frac{1}{k^2} \approx 1.631689.^{22}$$

For a second example, we will examine a sequence function that is defined recursively, that is, defined in terms of itself. Observe in Figure I.15 that we have entered the function $u1(n) = \cos(u1(n-1))$ and $ui1 = 0$. The first few terms of the sequence are, therefore: 0, 1, cos 1, cos (cos (1)), cos (cos (cos (1))), and so on.[23] Figure I.16 depicts the graph and the WINDOW parameters used to get the graph.

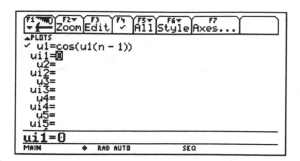 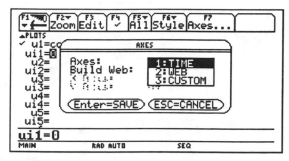

FIGURE I.15.

The sequence is entered (left). Be sure that "1:TIME" is chosen under the [F7] "Axes..." option (right).

---

[22]It can be shown in calculus that the *exact* value of $\sum_{k=1}^{\infty} \frac{1}{k^2}$ is $\frac{\pi^2}{6} \approx 1.64493406685$.

[23]The numerical values of the sequence can be easily evaluated in the HOME screen using the "seq(" function. Some patterns are shown below in FLOAT3 mode:

The top line above shows the first seven iterations of the sequence to three decimal places

To three decimal places, there appears to be no change in the values of the sequence after the 21st term.

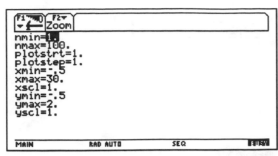
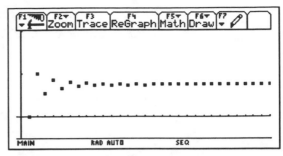

FIGURE I.16.

The terms of the sequence appear to be approaching a particular value. (Note that the style is automatically SQUARE.)

To approximate the value to which the terms seem to be converging, we use the $\boxed{F3}$ TRACE feature. Figure I.17 shows that value to be about 0.739085

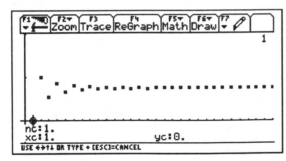
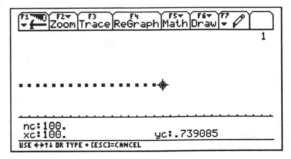

FIGURE I.17.

With TRACE, we enter 100 for the 100th term ("nc:") and see that the 99th iteration of this sequence is approximately 0.739085. (Note that $\boxed{\text{ENTER}}$ must be pressed again after the 100th term has been calculated.)

We conclude this section by showing one additional variation of sequence graphing -- a WEB plot. A WEB plot graphs the sequence by using $u1(n)$ values for the vertical axis and $u1(n-1)$ values for the horizontal axis. If a sequence converges to one specific value, as we have in our example, the WEB plot should converge to that point. The visual effect is very interesting.

To enter a WEB plot, we must choose "$\boxed{F7}$ Axes" from the Y = editor. For "Axes:" choose WEB and for "Build Web:" choose TRACE. (The other option here is AUTO mode and the WEB gets drawn automatically on the graph screen. We choose to have the WEB drawn from TRACE mode.) The "$\boxed{F7}$ Axes" screen, the WINDOW parameters, and the graph are displayed in Figure I.18.

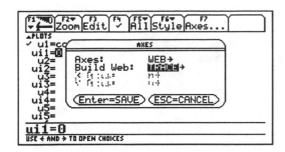

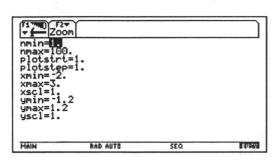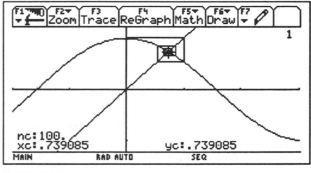

FIGURE I.18.

A WEB plot of $u1(n) = \cos{(u1(n-1))}$.
The WEB is constructed as you TRACE by repeatedly pressing the right cursor key.

I.4. 3D Graphing

The final graph type we will examine is 3D graphing -- for graphing a function of two independent variables, $z = f(x, y)$. Of course, like the other graph types, choose the 3D graph type in the first category of the MODE menu of the ***TI-92***.

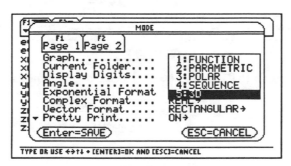

FIGURE I.19.

Selecting 3D graphing mode.

It should be obvious that when graphing in 3D mode, there will be additional WINDOW settings to consider. (Actually, "WINDOW" settings now refer to a viewing **cube** rather than a viewing rectangle.) Figure I.20 displays the default WINDOW values. Keep in mind that these are the values that come as a result of a ZoomStd in 3D mode.

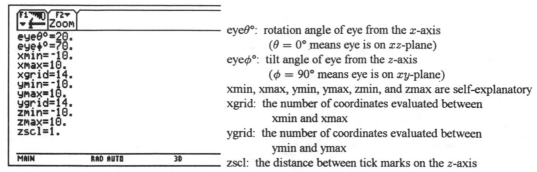

eye$\theta°$: rotation angle of eye from the x-axis
    ($\theta = 0°$ means eye is on xz-plane)
eye$\phi°$: tilt angle of eye from the z-axis
    ($\phi = 90°$ means eye is on xy-plane)
xmin, xmax, ymin, ymax, zmin, and zmax are self-explanatory
xgrid: the number of coordinates evaluated between
      xmin and xmax
ygrid: the number of coordinates evaluated between
      ymin and ymax
zscl: the distance between tick marks on the z-axis

FIGURE I.20.
The ZoomStd (default) WINDOW parameters for 3D graphing.

For a first example of a graph of a function of two independent variables, consider $z = 10e^{-x^2-0.5y^2}$. Figure I.21 shows that we have entered the function as $z1$ in the Y = editor. Also, it is important to choose the format of a 3D graph. (Recall, the format of any graph is determined by making choices from the "9:Format ..." option under [F1] menu from the Y = editor screen or by pressing [◇] [F].)

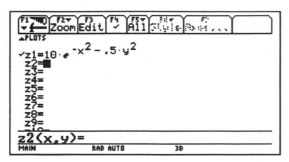 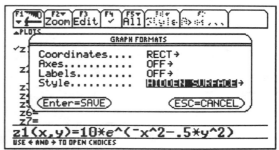

FIGURE I.21.
The Y = editor (left) and the [◇] [F] menu for graph formats (right).

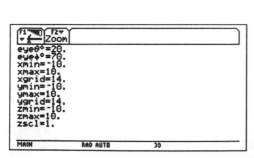 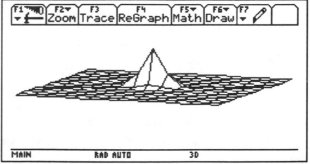

FIGURE I.22.
The graph of $z = 10\,e^{-x^2-0.5y^2}$ viewed with the default settings.
Be prepared to wait while 3D graphs are being drawn!

It is interesting to observe a 3D graph from a variety of viewpoints. For one variation of the graph in Figure I.21, we change ϕ to be 110°. This means that the eye is viewing the figure from 20° *below* the xy-plane. (If eye$\phi°=90$, the viewpoint is from the xy-plane.) See Figure I.23.

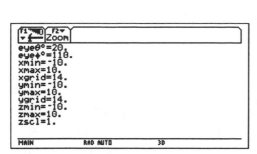

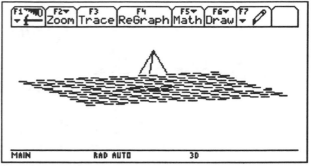

FIGURE I.23.

The graph of $z=10\,e^{-x^2-0.5y^2}$ viewed from an angle of 20° below the xy-plane. ($\phi=110°$.)

A final view of this same function appears in Figure I.24. Here we added the "box" into which a 3D graph is contained and also labeled the axes. These were added from the ◇ F menu (which can also be accessed by pressing F1 "9:Format").

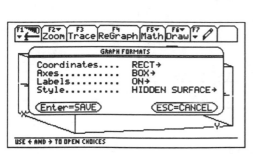

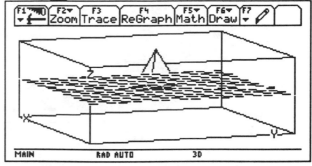

FIGURE I.24.

The graph of $z=10\,e^{-x^2-0.5y^2}$ viewed within the graphing WINDOW box (with Labels on).

J. The Calculus Menu

There are eleven (11) calculus commands and they can be accessed by either pressing the F3 key from the home screen or choosing "A:Calculus" from the 2nd MATH key. These menus are displayed in Figure J.1.

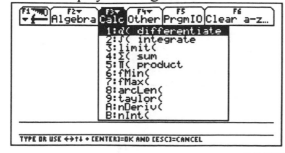

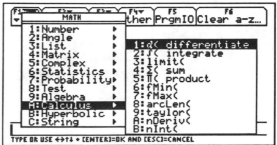

FIGURE J.1.

The calculus menu is accessed in the HOME Screen by F3 (left)

or by pressing A from the 2nd MATH key (right).

We will illustrate each of the eleven available commands in the following pages.

J.1. d(differentiate

The syntax for this command[24] is:
$$\texttt{EXPR,VAR[,ORDER]}$$

This means that after choosing "1:$d($ differentiate" from the calculus menu[25], an expression followed by a comma followed by the variable you are differentiating with respect to must follow. Also, the order of the derivative is an optional entry. For example, to find the first derivative of $(x^3 - 2x + 1)^2$ with respect to x, $\dfrac{d(x^3 - 2x + 1)^2}{dx}$, we enter the expression as highlighted in Figure J.2 below.

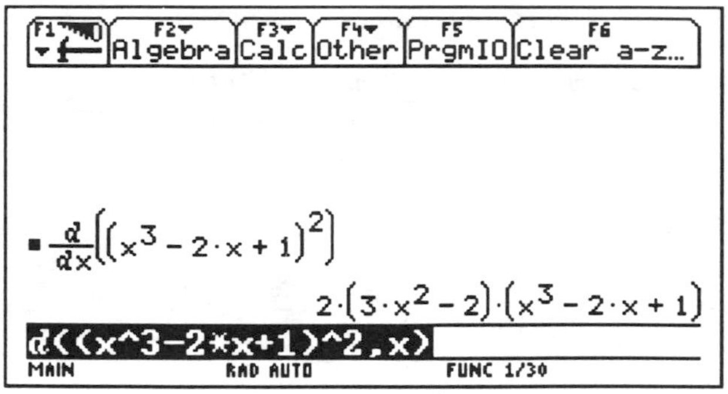

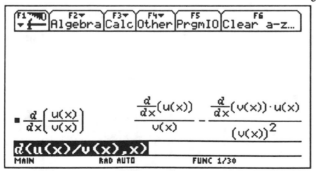

FIGURE J.2.

The first derivative of $(x^3 - 2x + 1)^2$ with respect to x is $2(3x^2 - 2) \cdot (x^3 - 2x + 1)$.

The symbolic capability of the *TI-92* even allows you to enter functions of x. For example, Figure J.3 displays the Quotient Rule for finding the first derivative of $\dfrac{u(x)}{v(x)}$.

FIGURE J.3.
The Quotient Rule.

---

[24]This syntax can be seen for any command or function from the 2nd CATALOG key. It appears in very tiny letters at the bottom of the CATALOG screen for the chosen command.
[25]Alternatively, "$d($" can be entered directly from the keyboard as the 2nd function of the numeric 8 key.

Of course, if the form of the answer in Figure J.3 does not match the form in your textbook, you can try using the FACTOR command on the result. (If you get an error message indicating that "Name is not a function or program", that may mean that either u or v has been previously define or used for some other purpose which interferes with the ability of the ***TI-92*** to differentiate it. You may need to use the "DelVar" command from the ▣ Other menu. For example, DelVar u, v would make u and v available for function names to be differentiated.) See Figure J.4.

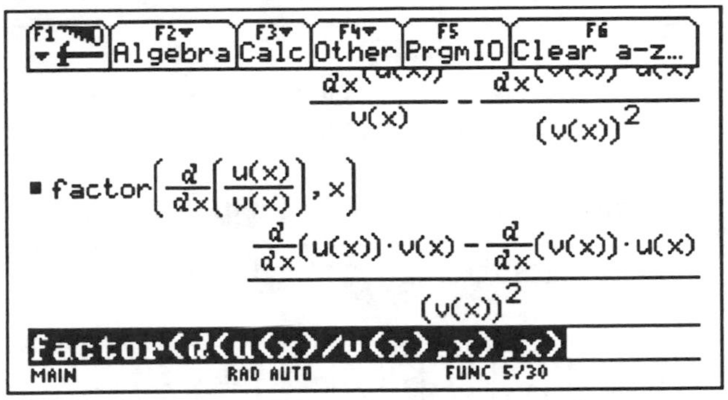

FIGURE J.4.
The Quotient Rule revisited.

We conclude this brief discussion of differentiation with two more examples. Figure J.5 displays the syntax for a second order derivative, namely, $\dfrac{d^2(\sin(x^2))}{dx^2}$.

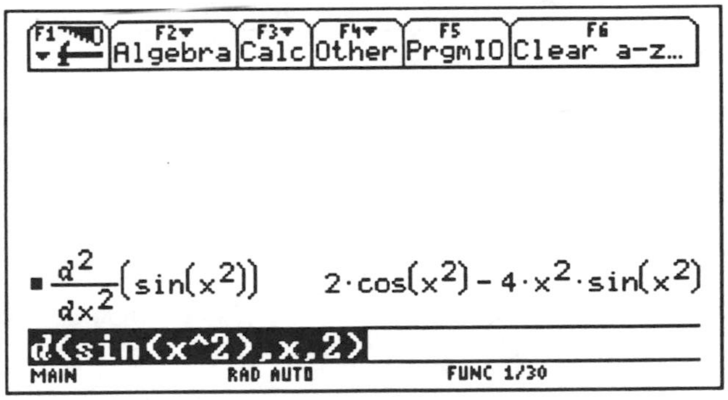

FIGURE J.5.
The second derivative of $\sin(x^2)$ with respect to x.

It should be noted that the "$d($" derivative option can be useful when graphing a function and its derivative. For example, consider the function $f(x) = x^3 + x^2 - 3x - 3$. If we enter it as $y1(x)$ and its derivative as $y2(x)$ (as in Figure J.6), we can observe the simultaneous behavior of a function and its derivative. Furthermore, if $y1(x)$ were to change (in another example) there would be no need to change the syntax for $y2(x)$.

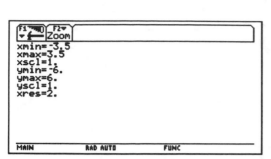 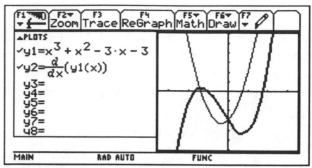

FIGURE J.6.

Notice that the function ($y1$ in thick style) is increasing on the exact intervals where its derivative ($y2$ in thin style) is positive.

Finally, we mention that the *TI-92* can be useful in finding partial derivatives. If $f(x, y, t) = x \cdot \cos(t) + y \cdot \sin(t)$ then $\dfrac{\partial f}{\partial t}$ is found in Figure J.7.

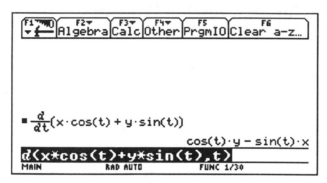

FIGURE J.7.

$\frac{\partial f}{\partial t} = y \cdot \cos(t) - x \cdot \sin(t)$ where $f(x, y, t) = x \cdot \cos(t) + y \cdot \sin(t)$

J.2. $\int ($ *integrate*[26]

The syntax for this command is: EXPR,VAR[,LOW,UP]

Here, the "low" and "up" are optional and used for definite integrals. If they are omitted, the *TI-92* will attempt to find the *indefinite* integral, that is the antiderivative of the integrand. (But note that the general constant of integration, " $+ C$ ", is omitted from evaluations on the *TI-92* unless you use the syntax $\int (f(x), x, c)$, which we do not recommend.) Figure J.8 shows two variations for the integrand expression x^2 :

---

[26]Alternatively, " $\int ($ " can be entered directly from the keyboard as the 2nd function of the numeric 7 key.

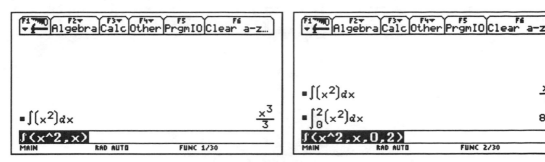

FIGURE J.8.
Compare the syntax of the highlighted entries. On the left we have the indefinite integral $\int x^2\,dx$
and on the right the definite integral $\int_0^2 x^2\,dx$.

Multiple integrals can be evaluated as well. This is accomplished by embedding integral symbols as can be seen in the highlighted entry in Figure J.9.

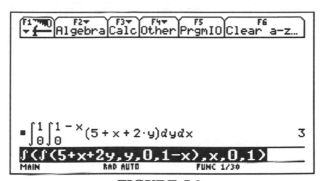

FIGURE J.9.
Evaluating a double integral.

J.3. limit(

The syntax for this command is: `EXPR,VAR,POINT[,DIRECTION]`

"`DIRECTION`" is for one-sided limits. Any negative value for `DIRECTION` is for calculating a limit from the left (i.e., limit from below) and any positive value for `DIRECTION` is for calculating a limit from the right (i.e., limit from above). If 0 or no value is entered for `DIRECTION`, the *TI-92* will calculate a two-sided limit.

For example, to find $\lim\limits_{x\to 2}\dfrac{x^2-4}{x-2}$, we enter limit((x^2 − 4)/(x − 2),x,2), the highlighted expression in Figure J.10. The value of 4 is returned.

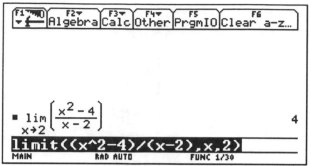

FIGURE J.10.
Evaluating a two-sided limit.

To evaluate $\lim\limits_{x \to 3^+} \dfrac{1}{x - 3}$, we enter limit(1/(x − 3),x,3,1); the *TI-92* returns ∞. See Figure J.11.

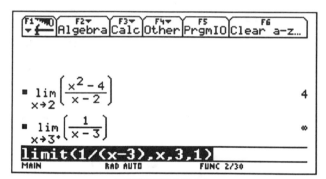

FIGURE J.11.
Evaluating a limit from the right.

Finally, a limit from below, for the greatest integer function (entered as "floor" on the *TI-92*), is calculated in Figure J.12 below.

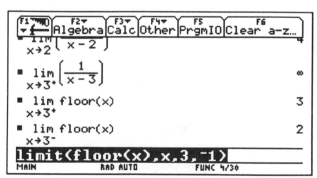

FIGURE J.12.
Evaluating a limit from the left.

J.4. Σ(sum and Π(product

The syntax for each of these commands is similar and we cover them together here. They are used for finding the finite or infinite sum or product of an expression. In either case, the syntax is:

$$\texttt{EXPR,VAR,LOW,HIGH}$$

So, to find $\sum_{n=1}^{\infty} \dfrac{1}{n^4}$, for example, we choose "4:Σ(sum" from the ⌷F3⌷ Calc menu and follow it

by typing 1/n^4,n,1,∞). To find $\prod_{t=1}^{5} \dfrac{1}{t^2}$ we choose "5:Π(product" from the ⌷F3⌷ Calc menu

and then enter 1/t^2,t,1,5). See Figure J.13.

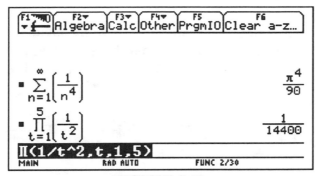

FIGURE J.13.

Evaluating an infinite sum and a finite product.

J.5. fMin(and fMax(

Each of these commands has similar syntax -- the user supplies two arguments, `EXPR,VAR` within parentheses. fMin(is used to find the minimum value of a function and fMax(is for finding the maximum value of a function. The *TI-92* returns possible x values that minimize (or maximize) the function.

For example, consider the $f(x) = 2x^4 + x^3 - 7x^2 - 4x - 4$. By observing its graph, we see that this function has two relative minima and one relative maximum.[27] We employ the "fMin(" option in Figure J.14, below.

---

[27]Of course, it is essential to have some idea of the function's behavior. The strategies we employ in this section assume we have studied the graph of the function:

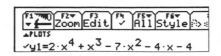

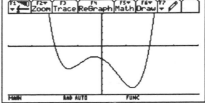

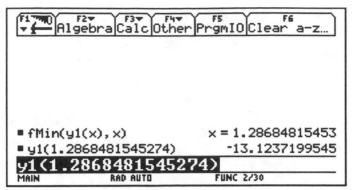

FIGURE J.14.
The (absolute) minimum of this function is about -13.1237 and occurs at $x \approx 1.28685$.

To find the other relative minimum value, we use the "with" operator. We append "$|x < 0$" to the "fMin(" command because we observe from the graph that another relative minimum appears to lie between $x = -2$ and $x = -1$. The "with" operator is obtained by typing $\boxed{2^{\text{nd}}}$ $\boxed{\text{K}}$ directly from the keyboard. See Figure J.15.

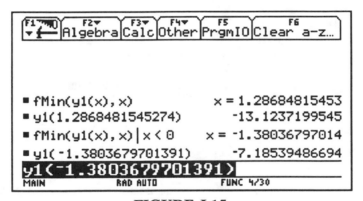

FIGURE J.15.
A relative minimum is about -7.18539 and occurs at $x \approx -1.38037$.[28]

To approximate the one apparent relative maximum point, we could try entering fMax(f(x),x) but the *TI-92* returns "$x = \infty$ or $x = -\infty$", since evidently it was in search of absolute maxima. Again, we employ the "with" option, this time using "$|x < 1$ and $x > -1$" as in Figure J.16 below.

---

[28]The *TI-92* displays the warning: "questionable accuracy" as a result of finding the relative minimum x value. This is probably due to the algorithm used to find the value. We have checked the values on this screen and they are reliable.

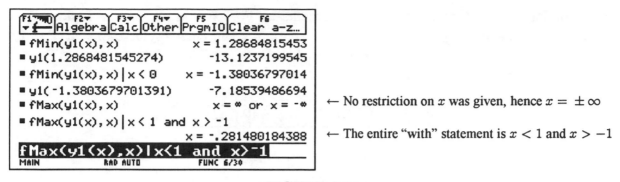

← No restriction on x was given, hence $x = \pm\infty$

← The entire "with" statement is $x < 1$ and $x > -1$

FIGURE J.16.
A relative maximum occurs at $x \approx -0.28148$.

J.6. arcLen(

The **TI-92** will calculate the length of an arc by choosing "8:arcLen(" from the F3 Calc menu. The user must then input

EXPR,VAR,START,END

where EXPR is a function of one independent variable. (Here, polar functions cannot be used.)

For example, to calculate the length of the perimeter of the unit *semi*-circle, $y = \sqrt{1-x^2}$ we enter arcLen($\sqrt{1-x^2}$, x, -1, 1). To find the length of arc of half the period of the sine curve, we would enter arcLen($\sin(x)$, x, 0, π). Both of these examples appear in the screen below:

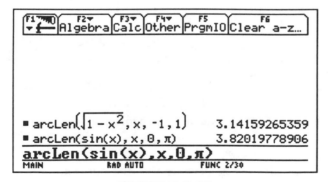

FIGURE J.17.
Two arc length examples (in FLOAT mode).

J.7. taylor(

To find the Taylor polynomial approximation to a function, we choose "9:taylor(" from the F3 Calc menu. The remaining syntax entered by the user should be EXPR, VAR, ORDER, and an optional POINT followed by a right parenthesis. For example, to find the Taylor polynomial approximation to $y = e^x$ of order 5 at $x = 0$, we can enter **taylor(e^x, x, 5, 0)** or, because the expansion point is 0, **taylor(e^x, x, 5)**. Both the fifth and seventh order Taylor polynomial approximations are depicted in Figure J.18 below.

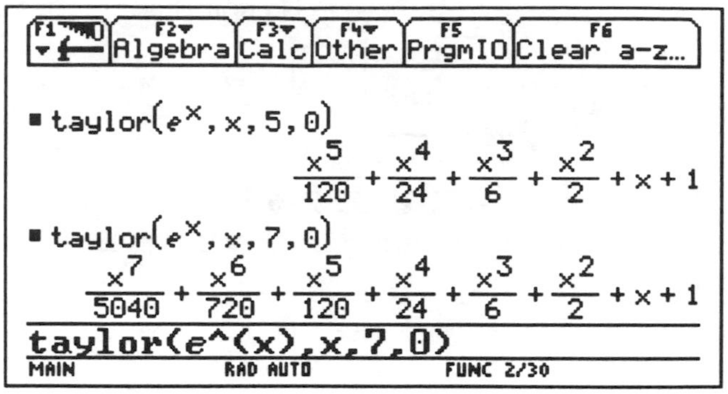

FIGURE J.18.

The fifth and seventh degree Taylor polynomial approximations to $y = e^x$ at $x = 0$.

If the expansion point is not zero, say it is $x = 1$, the *TI-92* will express the polynomial in powers of $(x - 1)$ -- probably the way the user would want it. If an expanded version were desired, you can use the "expand(" function from the [F2] algebra menu. We do this in Figure J.19.

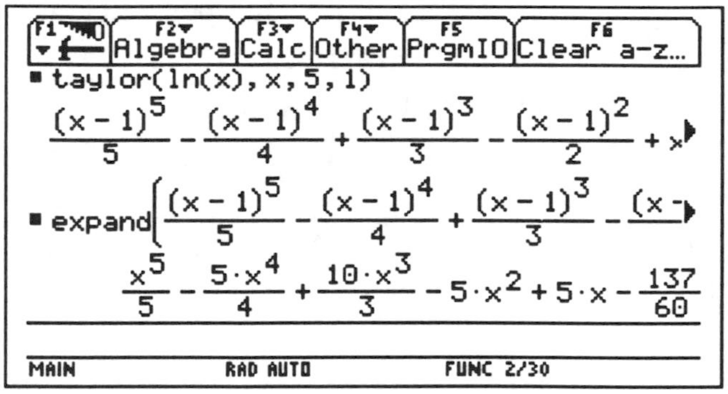

FIGURE J.19.

The fifth degree Taylor polynomial approximation to $y = \ln x$ at $x = 1$.

We complete this section by mentioning that the approximation by Taylor polynomials can be visualized nicely on the *TI-92* graphics screen. In Figure J.20 the function $y = e^x$ and its fifth degree Taylor polynomial approximation are graphed on the same screen.

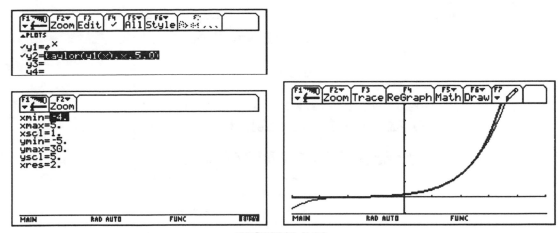

FIGURE J.20.

The thick graph is $y = e^x$; the thin line (from the "6:Path" style) is the 5th degree polynomial approximation.

J.8. nDeriv(

The "nDeriv(" option uses the central difference quotient formula[29] $\dfrac{f(x+h) - f(x-h)}{2h}$ to approximate the numerical value of the first derivative at x, $f'(x)$. We choose "A:nDeriv(" from the ▢F3 Calc menu. The syntax that follows the left parenthesis is: E X P R , V A R , H followed by a right parenthesis. The "H" is optional and defaults to 0.001 if it is not entered.

In the two screens that follow, we see a symbolic use of nDeriv(to calculate the derivative of the sine function (Figure J.21) and to approximate the first derivative of $y = x^4$ (in Figure J.22). Be sure that the variable h is not currently in use. If so, use DelVar to delete it.

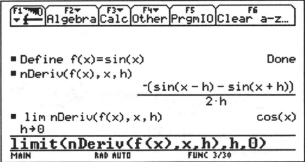

FIGURE J.21.

Using nDeriv and limit, we see that the derivative of the sine function is the cosine function.

---

[29]You may be used to the standard *difference quotient* (sometimes called the *forward difference quotient*) that appears in most textbooks: $\dfrac{f(x+h) - f(x)}{h}$. This quotient is built into the *TI-92* as "avgRC($f(x), x, h$)". If $f(x) = x^4$, we have:

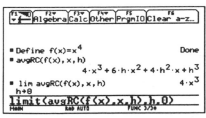

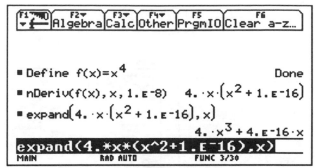

FIGURE J.22.

With $h = 10^{-8}$, the first derivative of $y = x^4$ is approximately $4x^3$.

K. Matrices and Systems of Equations

Matrices are entered from the keyboard using the square bracket keys. To enter the matrix
$\begin{bmatrix} 1 & 2 & 3 \\ -4 & 0 & 2 \\ 3 & 1 & -5 \end{bmatrix}$ and store it in location **a**, we enter $[[1, 2, 3][-4, 0, 2][3, 1, -5]]$ then press
the $\boxed{\text{STO} \triangleright}$ key followed by the $\boxed{\text{a}}$ key and then $\boxed{\text{ENTER}}$. See Figure K.1.

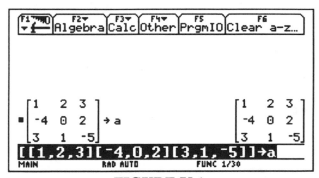

FIGURE K.1.

Entering a matrix and storing it in the variable **a**.

Individual components of matrix **a** can be accessed or changed by using **a**[m,n] which addresses the component in the mth row and the nth column of **a**. For example to change the element in the first row, third column of **a** from 3 to 4, we would enter 4 $\boxed{\text{STO} \triangleright}$ **a**[1, 3] as we see in Figure K.2.

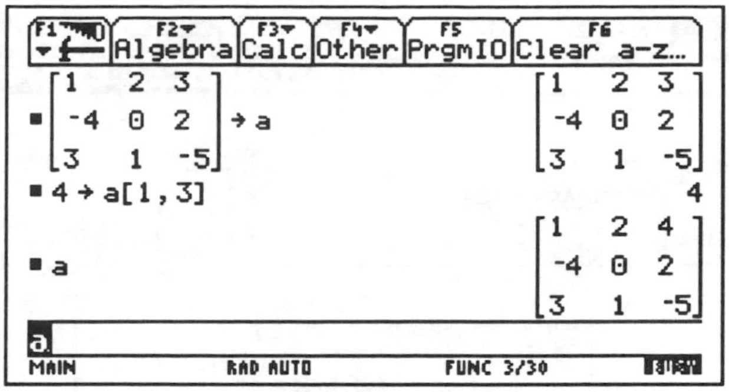

FIGURE K.2.

Changing one component of a matrix.

K.1. Matrix Operations

There are many matrix-related functions available to the *TI-92* user. They are available in the "4:Matrix" option under the MATH key. There are fifteen major menu items (labeled 1 through F in Figure K.3) and there are also many submenu items. At first, you may find the use of submenus a bit awkward when it comes to keeping track of options. But once you are familiar with the capabilities of matrix algebra on the *TI-92*, you will probably find the CATALOG key the most convenient way to locate a desired matrix function. See Figures K.3 and K.4 for an entire list of the matrix commands and functions.

FIGURE K.3.

A list of the fifteen main items available from "4:Matrix" of the MATH key.
(Notice that only 1 through C are initially available; you have to scroll down to see D, E, and F.)

FIGURE K.4.

The MATRIX menu items "B:Norms", "C:Dimensions", "D:Row ops", "E:Element ops", and "F:Vectors ops" each have submenu listings and these all appear here.

In the next few pages we will demonstrate some of these matrix functions. We return to the matrix **a** we entered earlier. One way to find the inverse of matrix **a** is to augment matrix **a** with the 3-by-3 identity matrix and then perform Gauss-Jordan row reductions on that augmented matrix. We use the *TI-92* "augment" command; it is highlighted in Figure K.5 below.

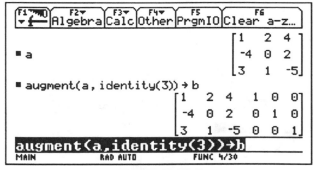

FIGURE K.5.

Here, we augment matrix **a** with the 3-by-3 identity matrix and store that result in (the 3-by-6) matrix **b**.

In Figure K.6, we find the reduced row echelon form of matrix **b** with the rref(b) function.

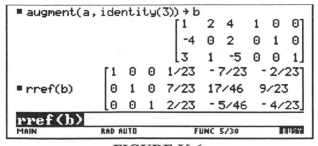

FIGURE K.6.

The reduced row echelon form of matrix **b**.

The inverse of matrix **a** could be found by entering a^−1 directly.[30]

Many operations on matrices are easily performed on the ***TI-92***. In Figures K.7 and K.8, the operations of matrix multiplication and addition are shown as well as multiplication of a matrix by a scalar.

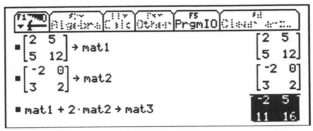

FIGURE K.7.

Scalar multiplication and matrix addition.

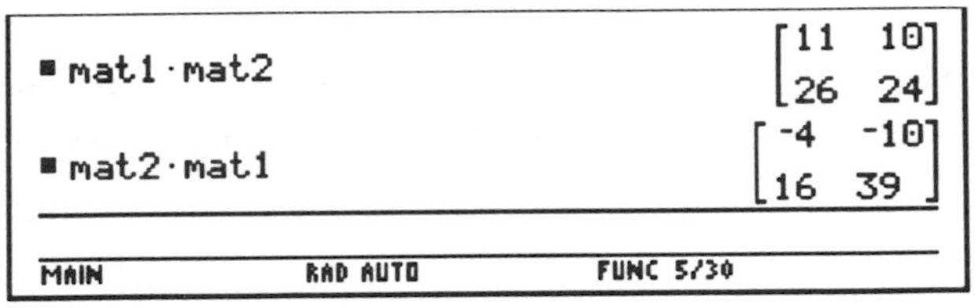

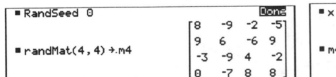

FIGURE K.8.

Matrix multiplication is not commutative! Notice that mat1*mat2 ≠ mat2*mat1.

Of course matrices need not be real-valued. Complex components are acceptable as are expressions and symbolic entries. (See Figures K.9 and K.10.) To generate an m-by-n matrix with random components, we use the "A:randMat(" option from the matrix menu. the entries will be randomly chosen from the integers between −9 and 9.

FIGURE K.9.

randMat(4,4) (left) creates a 4-by-4 matrix with random[31] entries; it is stored in matrix named m4. On the right, we place a different expression (say, "x") into the entry in the 3rd row, 4th column of m4 by typing x STO▷ m4[3, 4].

---

[30]The screen to find a^{-1} is:

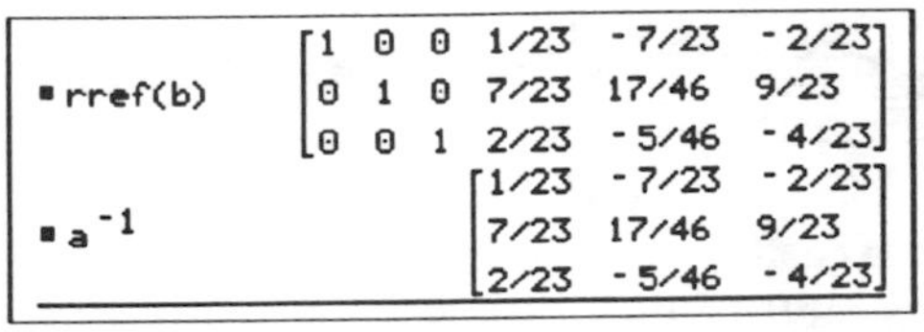

Also, the rref(b) function returns the ***reduced*** row echelon form of matrix **b**. The function ref(b) returns the row echelon form of **b**.

[31]We precede the "randMat(" command with *RandSeed 0* (which returns the random number generator to a fixed initial value) so that you can duplicate the screen we show here. Without the RandSeed 0 line, you may get different integers as entries.

```
F1▼  F2▼   F3▼  F4▼   F5    F6
▼⌐ Algebra Calc Other PrgmIO Clear a-z...
                              ┌ -3  -9   4   x ┐
                              └  0  -7   8   8 ┘
■ 2 - 3·i → m4[2, 1]                    2 - 3·i
                              ┌ 8       -9  -2  -5 ┐
                              │ 2 - 3·i  6  -6   9 │
■ m4                          │ -3      -9   4   x │
                              └ 0       -7   8   8 ┘
m4
MAIN        RAD AUTO       FUNC 6/30
```

FIGURE K.10.
We can have complex-valued matrices, too.

K.2. Systems of Linear Equations

There are a variety of ways to use the *TI-92* to solve a system of linear equations. We begin by considering the following (square) system of three equations in three unknowns:

$$x_1 + x_2 + x_3 = -1$$
$$3x_1 - x_2 + 5x_3 = 9$$
$$4x_1 + 2x_2 - x_3 = 2$$

We assign the matrix of coefficients $\begin{bmatrix} 1 & 1 & 1 \\ 3 & -1 & 5 \\ 4 & 2 & -1 \end{bmatrix}$ to the variable *a* and the column vector

of constants $\begin{bmatrix} -1 \\ 9 \\ 2 \end{bmatrix}$ to the variable *b*. Thus, our system can be written as $a \cdot x = b$ where x

is the column vector of unknowns, $x = \begin{bmatrix} x_1 \\ x_2 \\ x_3 \end{bmatrix}$. See Figure K.11.

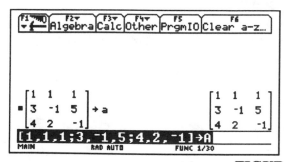
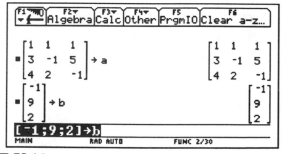

FIGURE K.11.
Storing the matrix of coefficients in *a* and the column vector of constants in *b*.
Notice the use of the semicolon to separate rows on the edit line.

Now, we choose the function "5:simult(" from the "4:Matrix" menu under $\boxed{2^{\text{nd}}}$ $\boxed{\text{MATH}}$ to solve the system. See Figure K.12 and observe that the solution is $x_1 = 2$, $x_2 = -3$, and $x_3 = 0$.

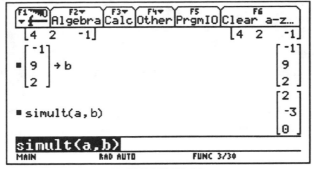

FIGURE K.12.

Solving the system $a \cdot x = b$ by entering simult(a,b) on the edit line.

Of course, to use simult(a, b), matrix a must be square and the length (the number of rows) of vector b must be the same as the number of rows of matrix a. Then, simult(a, b) returns a column vector of the same length.[32]   To solve a system that is not square, you would use the "rref(" operation on the appropriate augmented matrix.

L. Statistics

Statistics are entered and calculated by choosing "6:Data/Matrix Editor" from the $\boxed{\text{APPS}}$ menu. We will look at three examples in this section and the first one will involve one-variable statistical work. We open a new data variable called *tests*. See Figure L.1.

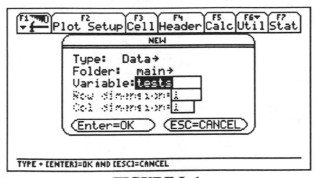

FIGURE L.1.

Beginning a new data file for statistics.

---

[32]In the event the coefficient matrix is singular (i.e., its determinant is zero), the following error message occurs:

L.1. Single Variable Statistics

Suppose Professor Plum had the following twenty scores on the first exam in his Calculus I course:

66, 74, 75, 82, 93, 58, 99, 79, 69, 62, 91, 90, 51, 66, 77, 88, 75, 75, 75, 82.

We enter these in column 1 ("c1") on the data screen as you can see in Figure L.2. (While editing, note that pressing the 2nd key before pressing the cursor up or cursor down key will jump a page at a time.)

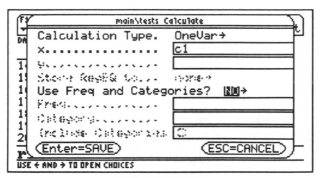

FIGURE L.2.
The twenty scores are entered in column c1. We also placed the reference word "Grades" at the top as a column title.

Now, to calculate statistics like the mean and standard deviation of these scores, we press the F5 key. In Figure L.3 we see the screen that appears when we press the F5 key to have the *TI-92* do these calculations.

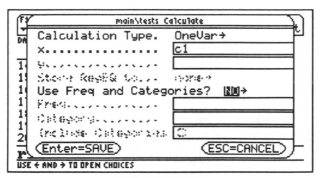

FIGURE L.3.
We choose "1:OneVar → " for calculation type and "c1" for the x-variable. Then press ENTER twice.

The results of these calculations are shown in the STAT VARS screen in Figure L.4.

```
┌────────────────────────────┐
│         STAT VARS          │
├────────────────────────────┤
│  x̄         =76.35          │
│  Σx        =1527.          │
│  Σx²       =119451.        │
│  Sx        =12.278672      │
│  nStat     =20.            │
│  minX      =51.            │
│  q1        =67.5           │
│  medStat▼75.               │
│                            │
│  q3        =85.            │
│  maxX      =99.            │
│    ( Enter=OK )            │
└────────────────────────────┘
```

What the statistics represent:

$\bar{x}$	the mean of the x values
$\sum x$	the sum of the x values
$\sum x^2$	the sum of the squares of x
Sx	the *sample* standard deviation[33]
nStat	the number of data points
minX	the minimum of the x values
q1	the first quartile
medStat	the median of x values
q3	the third quartile
maxX	the maximum of the x values

FIGURE L.4.

Ten statistics are displayed. (To see the last two, you will need to cursor down.)

Now press ENTER. Should you need to access this data (in the form of a list) later from the HOME screen, use the F1 "Save Copy As ..." option (also obtained by typing ◇ S.) See Figure L.5.

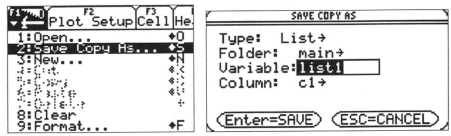

FIGURE L.5.

Here, we save the current data column c1 as a list that can be accessed from the HOME screen. Be sure to select "List" as the Type.

From the HOME screen, the "OneVar" command on the (newly created) list *list1* can be used to perform the one variable statistical calculations. Of course in order to access any of the values, we must enter their variable name on the EDIT line. In Figure L.6. we enter that command and then three different statistics.

---

[33]The standard deviation of the population is also calculated and is stored in memory location σx; it is not displayed here, however. It can be found by entering σx from the HOME screen. (The easiest way to do this is to enter the following keystrokes: 2ⁿᵈ G S X.)

```
 ■ OneVar list1                              Done
 ■ σx                                   11.967769
 ■ Sx                                   12.278672
 ■ x̄                                        76.35

 MAIN         RAD AUTO         FUNC 4/30
```

FIGURE L.6.
The population standard deviation, the sample standard deviation, and the mean are displayed in FLOAT 8
mode after the command OneVar is performed on the variable *list1*.[34]

L.2. *Frequency Distributions*

A manufacturer of car batteries has tested one type of battery to see how long it takes to
discharge when the electrical load is two normal headlights. A frequency table appears below
where the variable x represents the time in hours it takes the battery to discharge. You may
want to either delete old data[35] or begin a new data application.

x (in hours)	Freq
3.0	7
3.5	10
4.0	12
4.5	6
5.0	4
5.5	0
6.0	1

In order to enter this data into the *TI-92*, we need to open a new "6:Data/Matrix
Editor" file after pressing the APPS key. We call it *battery* and enter the x values in column
c1 and the frequencies in column c2. See Figure L.7.

---

[34]To display the mean of the x values on the HOME screen, select "2:Math" from the 2nd CHAR menu:

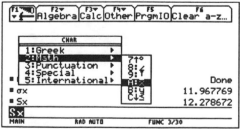

[35]To delete a previous column of data, move the cursor to the top of the column and press the F6 Util key and
select "2:Delete" followed by "3:Column." To clear all the data, select "8:Clear Editor" from the F1 menu
while in the data editor.

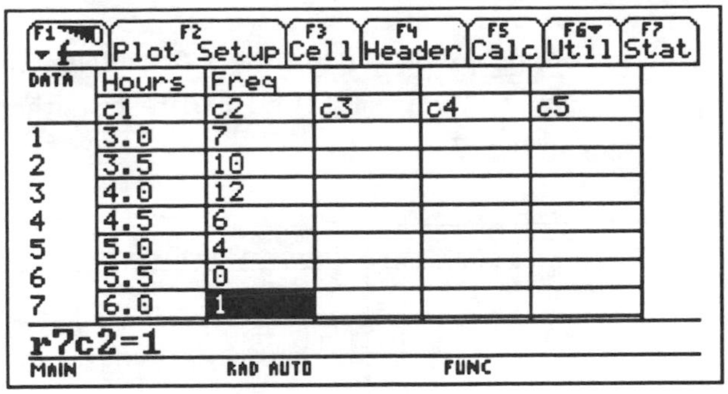

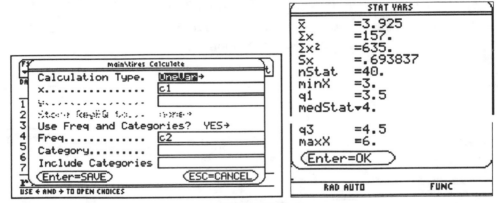

FIGURE L.7.

For this frequency table, the x values are entered in column c1 and the frequencies are in c2.
We are in a FIX 1 display digits mode.

Although we have entered two columns of numbers, we are still interested in one variable statistics (we really only have one variable: hours of battery operation). So, from the [F5] Calc menu we enter c1 for the x variable and c2 for frequency. See Figure L.8.

FIGURE L.8.

Notice that although the data was entered in seven rows, there are nStat = 40 pieces of data (the sum of the frequencies in c2).

Our next objective is to graph the data. The *TI-92* provides a variety of ways of doing that--we choose a histogram here. Other possibilities are scatter plot, an xy- line and a box plot. We choose [F2] Plot Setup and from that screen select [F1] to define the plot of this data. (There are nine possible plot definitions.) The plot definition screens are depicted in Figure L.9 below.

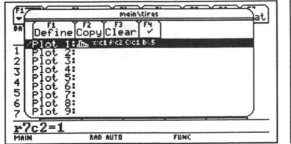

FIGURE L.9.

Defining a histogram plot.

After we choose a suitable graphing window from the ◇ WINDOW, we are ready to press ◇ GRAPH to see the histogram in Figure L.10. Be sure to either deselect or delete any other functions that may appear in the Y = editor.

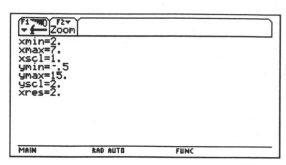 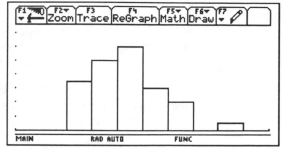

FIGURE L.10.
The histogram is drawn. Notice that ymin, ymax pertain to frequencies (column c2).

To adjust your histogram, you may want to do a ZoomData from the F2 Zoom menu. This will fill the screen from left to right with the histogram and center each vertical bar over the data point. Figure L.11 shows the result of a ZoomData.

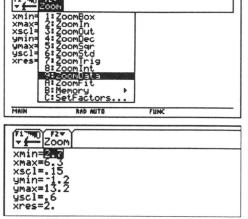 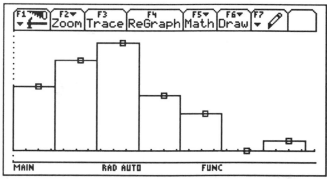

FIGURE L.11.
The result of a ZoomData on the previously draw histogram. Also included is a scatter plot (Plot 2:) of the data points.[36]

---

[36]The Plot Setup screen for the scatter plot appears below:

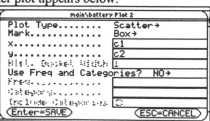

L.3. Two-Variable Statistics: Regression

To examine two-variable statistics on the *TI-92*, we use a model from business. When the Dow Jones Industrial Average (DJIA) broke the 5000 barrier on November 21, 1995, there was much speculation about when the next milestone (6000) would be reached. It is interesting to note that these milestones seem to be occurring at an exponential rate. They appear below:

Date	Coded Date (in years since 1/1/1906)	DJIA
	x	y
Nov. 14, 1972	66.9	1003
Jan. 8, 1987	81.0	2002
April 7, 1991	85.4	3004
Feb. 23, 1995	89.2	4003
Nov. 21, 1995	89.9	5000

To see this exponential growth, we enter the data as in Figure L.12 below along with a calculation of *exponential regression* (ExpReg) from the F5 Calc menu.

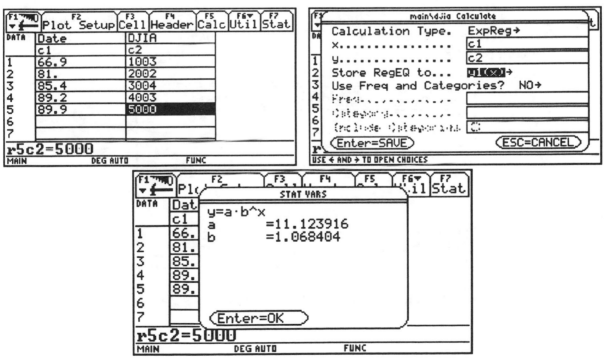

FIGURE L.12.

The calculation of exponential regression. Notice that the equation $y1(x) = a \cdot b^x$ will automatically be entered as $y1(x)$ in the Y = editor, overwriting any other function defined as $y1(x)$.

After choosing a suitable set of WINDOW values, and defining a scatter diagram of the points, we are ready to examine the points along with the exponential function of best fit. The resulting graph is displayed in Figure L.13.

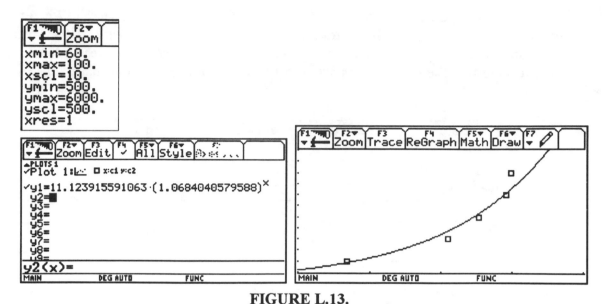

FIGURE L.13.
The graph of $y1(x) = a \cdot b^x$ with the plotted data points.

Using this function as a predictor along with the "solve(" command, the next milestone ($y1 = 6000$) should occur at $x \approx 95.0702$ or approximately, January, 2001. That calculation is done in Figure L.14 below.

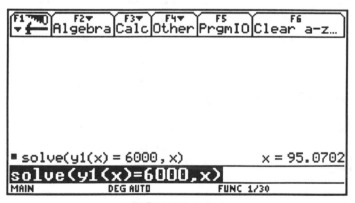

FIGURE L.14.
The projected date for the DJIA to hit 6000 is 95.0702 years $\approx$ January, 2001.

M. Geometry

The built-in Geometry Application is a powerful tool which enables the user to investigate and discover relationships, patterns and connections among mathematical concepts. It is completely interactive, encouraging explorations that begin with "what if". This section will lead the user through many, although not all, of its features. Note that some screens in this section might look slightly different than yours, particularly in the placement of the cursor. It may have been moved to unclutter the final screen.

The geometry application is accessed by pressing the [APPS] key and then selecting "8: Geometry" as in Figure M.1. Then, from the submenu, you have three choices: work on a **current** document, **open** a previously saved one or, begin a **new** document. Let's begin a new document now by typing [3] (or move the cursor to "3: New" and press [ENTER]).

 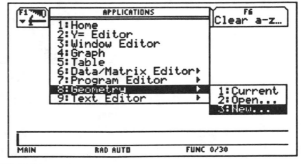

FIGURE M.1.

Opening a new geometry session.

A dialog box appears which asks for the name of this document. Move the cursor to the blank line and type *tri* and press [ENTER] twice. See Figure M.2.

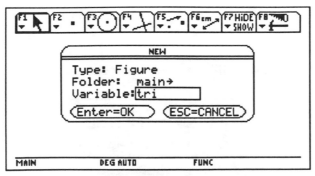

FIGURE M.2.

We name our first session *tri*.

Now, you see a blank screen with a toolbar menu across the top. This menu bar consists of eight groups of tools and operations, each containing a submenu. In this chapter we will examine some of the operations in each group. Each menu is accessed by one of the keys, [F1] through [F8], located on the left side of the calculator.

Important: When you activate one of the tools, it remains activated until another tool is chosen. The exception is the Macro tool under [F4], which will be discussed later, and the [F8] menu items.

M.1.   [F1] *The Pointer Menu*

Now, press [F1] and see the menu depicted in Figure M.3. The first tool "1: Pointer", is the tool you will return to frequently. **Note:** You can return to this tool easily at any time by pressing [ESC].

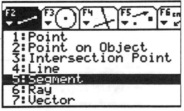

FIGURE M.3.
The [F1] pointer menu.

You usually will return here after using another tool, since the pointer is used whenever you want to select an object to move it, change its shape, or delete it. Before the other tools ("2:Rotate", "3:Dilate", and "4:Rotate & Dilate") on this menu can be used, we need an object upon which to operate. To *construct* such an object a different tool is needed. Let's press [ESC] to leave this menu now and proceed to [F2].

M.2.   [F2] *Points and Lines Menu*

Now, press [F2]. Notice that seven different tools are located under this submenu. Let's select "5: Segment."

```
F2        F3        F4        F5        F6
1:Point
2:Point on Object
3:Intersection Point
4:Line
5:Segment
6:Ray
7:Vector
```

FIGURE M.4.
The [F2] points and lines menu.

Note: After selection, the tool icon for [F2] on the tool bar changes from a point to a segment as a reminder of which tool is selected on the submenu. (The entire [F2] icon has a thick box around it to show it is selected.)

Since a segment is defined by two points, the points need to be positioned. Using the cursor pad which can move in eight directions, move the icon (now shaped like a pencil) to a point near the middle of the screen and press ⏎ENTER. Next, move the cursor toward the right side of the window, until the segment is of the desired length and press ⏎ENTER. The segment is now defined. See Figure M.5.

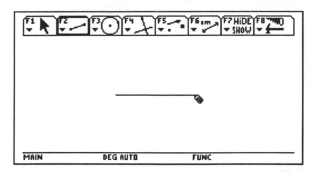
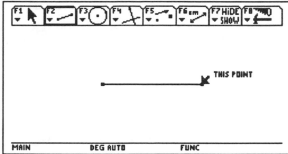

FIGURE M.5.
Drawing a line segment: position the first endpoint, press ⏎ENTER, position the second endpoint and press ⏎ENTER again.

Notice that the segment tool is still selected as you can see from the toolbar. If the cursor is moved to another point (or the final endpoint is used as an initial point) the process can be repeated to create another segment.

Note: In addition, a point may be labeled immediately after its creation by typing a letter after pressing ⏎ENTER to locate the point. Use the shift key (⇧)before a letter to type uppercase.

Now, press F1 and then "2: Rotate". Notice, the cursor has become a small + sign. Move it until it touches the segment and the words "this segment" appear. (The cursor becomes an arrow.) Press ⏎ENTER. The segment now becomes a (flashing) dotted line, indicating that it has been *selected* for some operation. See Figure M.6.

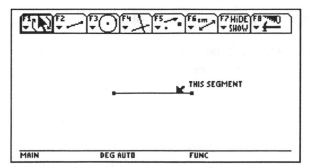
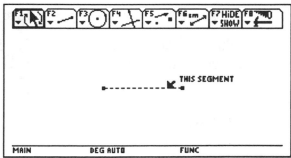

FIGURE M.6.
After choosing the rotate option (notice the F1 icon), we select the segment and it becomes a flashing dotted line.

Now, while holding down the hand key, 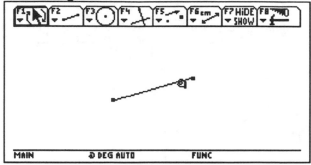 (above the F5 key), cursor-up until the segment begins to rotate. Until you press ENTER, the segment remains selected and can continue to be rotated as you can see in Figure M.7.

FIGURE M.7.
The rotated line segment.

Notice the sequence was to:
> 1) select the tool
> 2) select the object upon which to operate
> 3) operate on the object.

Until another tool is selected, the rotation tool remains selected.

While we are practicing, objects frequently will need to be deleted. The sequence will always be to select the object(s) and then delete. Objects can be selected in several ways.

To select one object:
> 1) Use the pointer tool. (Move to an object until it is named and press ENTER. It now is *selected* and becomes (flashing) dotted. Then, press the delete key, ⬅.)

or

> 2) Use the selection rectangle. (Move the pointer so that it is positioned above and to the left of the desired object. Hold down the hand key, 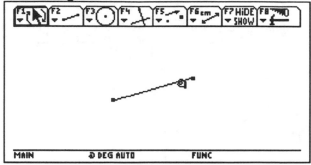, while moving the cursor to the lower right corner. This selects the object as the cursor moves across it. Then, press the delete key, ⬅.)

To select multiple objects:
> 1) After selecting the first object with the pointer, hold down the shift key while moving to another object, cursor up and press ENTER. Continue in this manner, using both keys to select as many objects as you wish. Then, press the delete key, ⬅.

or

> 2) Use the selection rectangle as above, dragging across multiple objects.

To clear the screen completely, press ⬚F8 and choose "8:Clear All". Then press ⬚ENTER when prompted. Let's do that now.

M.3. ⬚F3 *Curves and Polygons Menu*

This menu, depicted in Figure M.8, contains items to draw a variety of geometric shapes.

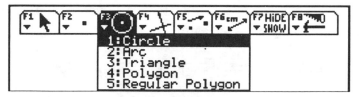

FIGURE M.8.
The ⬚F3 curves and polygons menu.

1: Circle

Let's begin by constructing a circle. Press the ⬚F3 and then type ⬚1 or press ⬚ENTER, since number one already is highlighted. A circle is defined by a center point and a radius so first we'll move the new cursor (the drawing pencil) to a position where the circle will be centered and press ⬚ENTER. Then, use the cursor pad to move the pencil until the circle is the desired size. After the radius is adjusted, press ⬚ENTER to complete the circle. See Figure M.9.

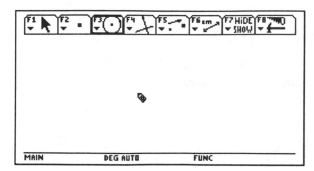 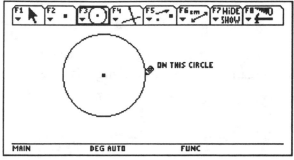

FIGURE M.9.

Constructing a circle: Left, after choosing "1:Circle" from the ⬚F3 menu, move the cursor to the place on the screen where you want the circle's center. Press ⬚ENTER. Then move the cursor until the desired radius is achieved and press ⬚ENTER (right).

To change the circle's size, you must select it before trying to make any adjustments. Press ⬚F1 and select the pointer tool. Then, move the pointer to the circle, until the words "this circle" appear and press ⬚ENTER to select the circle. Then while holding down the hand key, 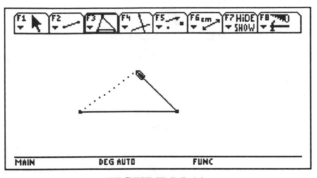 move the cursor until the circle is the desired size and press ⎡ENTER⎤. Notice that the center did not change--just the radius. Again, we selected an objected *before* performing an operation on it. Doing the same thing with the circle's center would move the entire circle.

3: Triangle

Let's create a different figure but first clear the entire screen. Now, press ⎡F3⎤, and select "3:Triangle". Three vertices are needed to define the triangle. Begin by positioning the cursor for the first vertex and press ⎡ENTER⎤. Then, move the cursor to the position of the second vertex and press ⎡ENTER⎤. As the cursor is moved to a third vertex, notice (Figure M.10) that a dotted line completing the triangle has appeared.

FIGURE M.10.
Constructing a triangle.

The dotted line will become solid when ⎡ENTER⎤ is pressed for the third vertex. Do that now.

Let's change the location of the triangle. Choose the pointer tool under ⎡F1⎤ and select the triangle. Now, hold down the hand 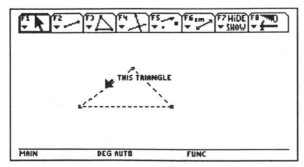 while moving the cursor to drag the triangle to a new position. When the triangle is where you want it to be, move the cursor away from the triangle and press ⎡ENTER⎤. See Figure M.11.

 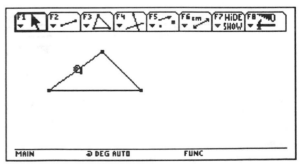

FIGURE M.11.
Selecting and moving a triangle.

Now, let's change the shape of the triangle. If the pointer tool ($\boxed{\text{F1}}$, $\boxed{1}$)is not still selected, select it now. Then, move the cursor to a vertex until the words "this point" appear. When $\boxed{\text{ENTER}}$ is pressed, that vertex will be selected. Now, while *holding down* 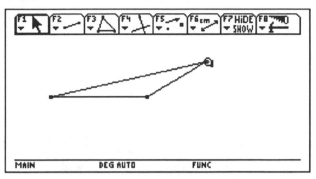, use the cursor to **drag the vertex** to a new position. If you are satisfied with the new shape, release 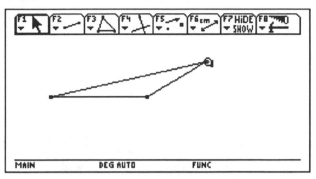, move the cursor off the figure and press $\boxed{\text{ENTER}}$.

FIGURE M.12.
Changing the shape of a triangle.

5: Regular Polygon

Let's continue our exploration. First, we need to remove the previous construction. Clear the screen using "8:Clear All" from the $\boxed{\text{F8}}$ menu. Another useful tool under $\boxed{\text{F3}}$ is "5:Regular Polygon". Choose that tool now.

FIGURE M.13.
Select "5:Regular Polygon" from the $\boxed{\text{F3}}$ menu.

The tool operates by drawing chords on a (circumscribed) circle. The cursor is a pencil and it is used to draw the circle. First, let's locate the center of the desired polygon and press $\boxed{\text{ENTER}}$. Then, move the cursor until the dotted circle is the size desired. Now, upon pressing $\boxed{\text{ENTER}}$, one chord of a polygon appears. There is a small number in the center, indicating the number of sides the polygon will have. The default number is 6 (see the left screen in Figure M.14). Move the cursor to the right (and/or up) to increase the number of sides, to the left (and/or down) to decrease that number. Notice how the number in the center of the circle is changing.

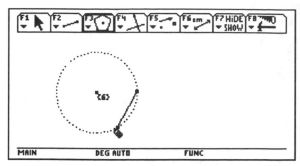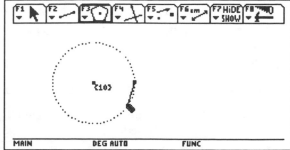

FIGURE M.14.

Left: Beginning a polygon (default is six sides).
Right: As you move the cursor up or to the right the number of sides increases.

For this exercise let's create a pentagon. Move the cursor until the number in the center is 5. Then, press [ENTER]. DO NOT press [ENTER] again, until the pointer tool ([F1], [1])is selected unless you want to create another polygon immediately. (See Figure M.15.)

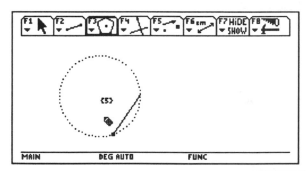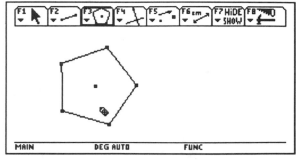

FIGURE M.15.

Constructing a regular pentagon.

M.4. [F4] Construction Menu

The [F4] key provides eight tools (the menu items in Figure M.16) that aid in the construction of a variety of special lines and curves. Included as well is a Macro tool which is used to define a sequence of construction steps that can be repeated easily.

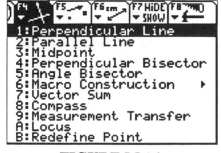

FIGURE M.16.

The eleven [F4] menu items.

To begin, clear the screen (by choosing "8:Clear All" from the F8 menu) and we'll draw a straight line. This is done by pressing F2, and then selecting "4:Line", the line tool. Two points are needed to define a line. First, move the cursor to any desired point which will be on the line and press ENTER. Now, further movement of the cursor generates a line whose direction depends on the position of the cursor. As the cursor is moved, the line's direction changes. Press ENTER when the line you want is drawn on the screen as we have done in Figure M.17.

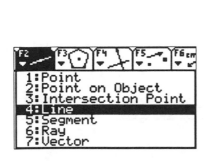
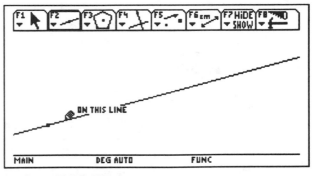

FIGURE M.17.

Drawing a line from the F2 menu.

1: Perpendicular Line

We can now construct lines parallel and perpendicular to this line. First, press F4 and choose "1: Perpendicular Line". If it is not already there, move the cursor to the line on the screen until the words "perpendicular to this line" appear. Then press ENTER. Move the cursor to the point on the line where the perpendicular line is desired ("on this line" appears) and press ENTER. A line, perpendicular to the original line, will immediately be drawn at that point. See Figure M.18.

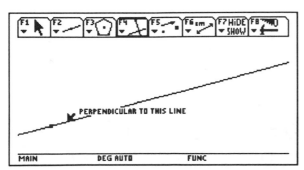
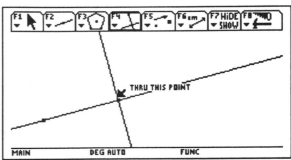

FIGURE M.18.
Drawing a line perpendicular to the original line (from Figure M.17).

Note: After the perpendicular line tool is selected, an initial line and desired point of intersection must be indicated. The order does not matter (nor does the location of the point).

2: Parallel Line

Now, let's continue exploring with the parallel line tool. From the [F4] menu choose "2:Parallel Line". As with other tools an *object* (here, an initial line) must be chosen. In addition, this tool requires selection of a point which is to lie on the parallel line. Again, the order does not matter. Now, move the cursor to the original line until the words "parallel to this line" appear. Press [ENTER].

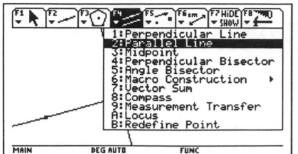 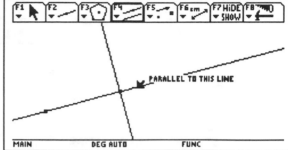

FIGURE M.19.
Drawing a line parallel to our original line: Step 1 is selecting that original line.

Then move the cursor to the desired location for the parallel line. The point which is to lie on the line can be chosen on the second line (the perpendicular we constructed), or anywhere on the screen. For this example, let's move the cursor to the open space above and to the right of the intersection of the existing lines. Now press [ENTER], and a line parallel to the original line, through the desired point will be drawn. See Figure M.20.

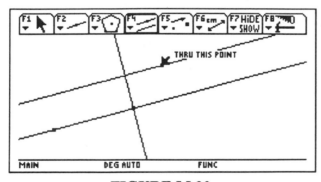

FIGURE M.20.
Completing the parallel line.

5: Angle Bisector

Next, let's work with the angle bisector tool. In order to use this tool an angle must be defined by three points, the second of which indicates the vertex of the angle. The following construction will create a triangle and use the angle bisector tool to locate the *incenter*, the center of the inscribed circle. First, let's clear the screen again (choose "8:Clear All" from the [F8] menu) and draw a triangle. After the triangle is drawn, press [F4] and then choose "5:Angle Bisector". See Figure M.21.

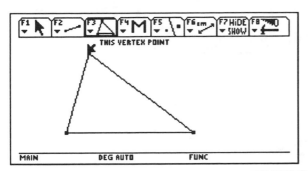 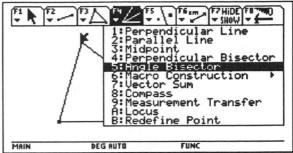

FIGURE M.21.

Draw a rectangle on a cleared screen and choose [F4] "5:Angle Bisector" for the work in this section.

Since the angle bisectors of a triangle are *concurrent* (that is, they all intersect at a single point), it will be sufficient to construct two of them, rather than all three. Let's begin by identifying one angle with three points, the middle point signifying the vertex of the angle. Move the cursor to one side of the angle until the words "on this triangle" appear. Press [ENTER]. One point is now identified. It will flash on and off until the bisector is finally drawn. Move the cursor to the vertex and the words "this point" appear. Press [ENTER] again and move the cursor along the second side to another point. As soon as you press [ENTER], the bisector is drawn. This four-step process is depicted in Figure M.22.

Step 1: Select one side of the triangle:

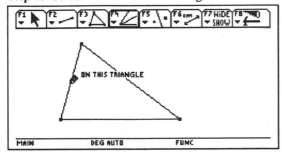

Step 2: Select the angle to be bisected:

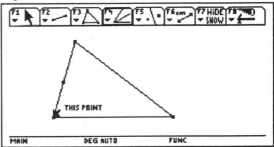

Step 3: Select the second side:

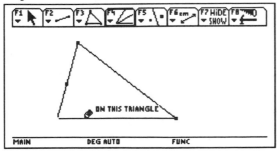

Step 4: Press [ENTER] and the angle bisector is drawn:

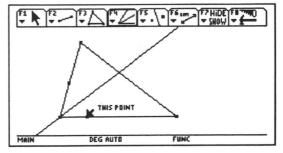

FIGURE M.22.
The angle bisector is drawn after the two second side is selected.

Now, three points are needed to define the other base vertex. Repeat the previous process. Figure M.23 shows the second angle bisector just after pressing [ENTER] for the third point.

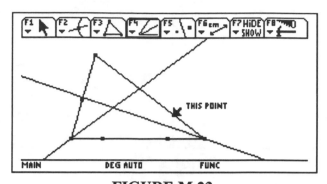

FIGURE M.23.

The second angle bisector is drawn.

This construction will be used as the foundation for the construction of the triangle's inscribed circle. First we need to identify the point of intersection of the bisectors. Press F2 and choose "3: Intersection Point". Move the cursor near the intersection of the bisectors until the words "point at this intersection" appear as in Figure M.24.

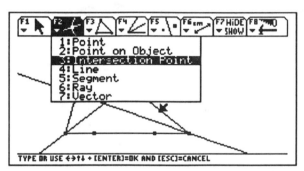
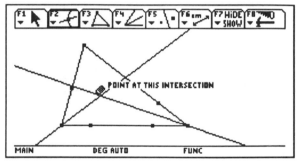

FIGURE M.24.

Select the point of intersection of the two bisectors.

Press ENTER .[37] Alternatively, you can move the cursor to one of the bisector lines and press ENTER. The line becomes dotted. Then move to the other bisector and press ENTER and the intersection point is drawn.

Next, we'll construct the inscribed circle. The incenter (center of the inscribed circle) has been located at the intersection of the angle bisectors, and the radius will be the perpendicular distance to one side. Therefore, we must construct a perpendicular to the base, through the incenter. We'll use the perpendicular line tool under F4 to do this. After selecting the perpendicular line tool, position the cursor at the intersection of the angle bisectors. Then press ENTER and move the cursor to the base. The words "perpendicular to this side of the triangle" appear. As soon as you press ENTER, the perpendicular is drawn. See Figure M.25.

---

[37]Typing a letter from the keyboard, say "a", will label this point if so desired.

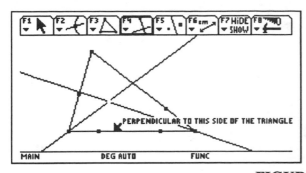 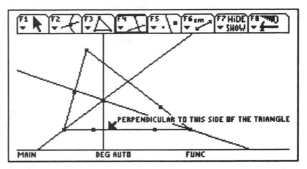

FIGURE M.25.
Drawing a perpendicular line through the angle bisector intersection.

A radius of our inscribed circle is the distance along this perpendicular from the point of intersection of the angle bisectors to the side of the triangle, since the incircle is tangent to each side. Now, choose the circle tool under F3. Move the cursor to the incenter until the words "this center point" appear. Press ENTER. Now move the cursor until the pencil is at the point of intersection of the base and the perpendicular bisector. It is very important that the pencil actually be at the point of intersection of the perpendicular and the base. It is not sufficient that the circle just appears tangent to the base. Press ENTER. The incircle is now complete. See below.

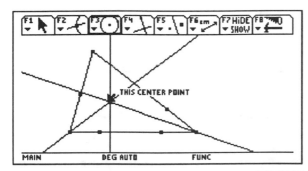 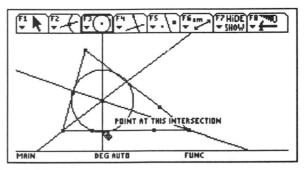

FIGURE M.26.
The incircle is completed.

It may appear, due to the shape of your original triangle, that we are assuming some special properties of the original triangle. To see that the construction relates to *any* triangle, grab a vertex (it will be easiest to grab the vertex we haven't used yet) and change the shape of the triangle. To do this, either press ESC or choose the pointer tool under F1. Then move it to the top vertex until the words "this point". Now, hold down the hand key, 🖐, while moving the cursor, dragging your original triangle to a different shape. See Figure M.27.

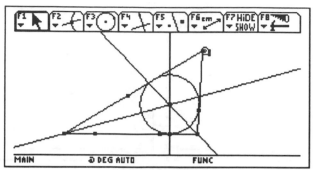

FIGURE M.27.

As the hand grabs and changes the original triangle, the constructions are updated based on the change!

Notice that all constructions that depend on the original triangle will be updated. The inscribed circle remains tangent to the three sides of the triangle no matter how the shape of the triangle is changed. The construction of the incircle assumed no special properties of the triangle--it will work for *any* triangle!

6: Macro Construction

The next tool we'll investigate is the *Macro* tool. Remember, a macro is a series of instructions that are repeated automatically to perform some task or construction. To write the macro, the entire construction is carried out first. Then, the macro will be defined based on the desired initial object(s) and the final object(s). When the construction is repeated automatically, any intermediate steps will not be shown.

The following steps create a macro that will **construct an equilateral triangle** on a given line segment. To begin, let's clear the screen (by choosing "8:Clear All" from the F8 menu). Using the segment tool under F2, draw a segment on the screen. For this exercise, make the segment rather short and locate it in the lower half of the screen. Next select the circle tool from F3. Position the cursor so the center of your circle is at the right endpoint of the segment. Then, move the cursor to the left until the pencil is at the other endpoint, indicating the segment is being used as the radius. The words "this radius point" appear. See Figure M.28.

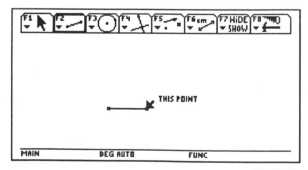 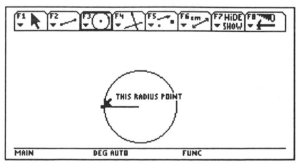

FIGURE M.28.

Left: Draw a segment (using the F2 menu). Right: Draw a circle (using the F3 menu) with the segment as its radius and right endpoint as its center.

Repeat the process using the left endpoint as the center of the next circle. We have constructed two overlapping circles with the same radius in Figure M.29. (If the upper point of intersection of the two circles is off the screen, either make the segment shorter by selecting and dragging one end of the segment or you can scroll the drawing window by pressing $\boxed{2^{nd}}$ and the hand key, ![hand key], at the same time. You may also refer to the directions for $\boxed{F8}$, "A:Show Page" later in this chapter.)

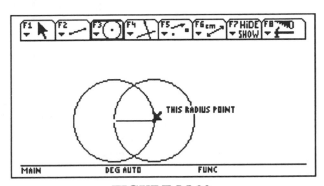

FIGURE M.29.
Constructing the second of two overlapping circles.

Press $\boxed{ENTER}$ to fix the second circle. Next, we'll identify a point of intersection of the two circles. Choose $\boxed{F2}$, "3: Intersection Point". Move the cursor to the top point of intersection until the words "point at this intersection" appear and press $\boxed{ENTER}$. See Figure M.30. Alternatively, you can click on each of the circles in turn with this tool and both points of intersection of the two circles will be defined.

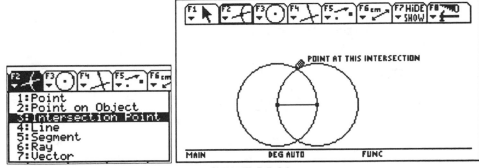

FIGURE M.30.
Our goal is to construct an equilateral triangle from the beginning segment. Defining the point of intersecting circles here will soon define the equilateral triangle since this point is on the perpendicular bisector of the original segment.

Now, draw two segments (select "5:Segment" from the $\boxed{F2}$ menu). Each segment will begin at the top point of intersection and will connect to one endpoint of the segment. The equilateral triangle has been constructed in Figure M.31.

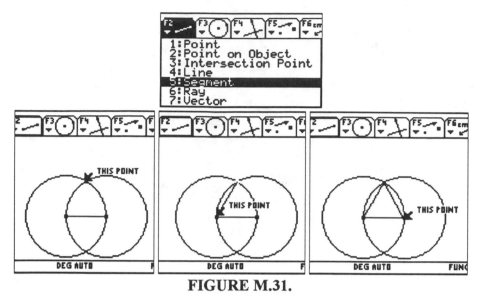

FIGURE M.31.
The equilateral triangle is completed by drawing the two new segments!

Now that the triangle has been constructed, the macro can be written. Press ⌊F4⌋, and highlight "6: Macro Construction". In the submenu choose "2: Initial Objects". Now, move the cursor to the initial segment. The words "this segment" appear. Now press ⌊ENTER⌋. See Figure M.32.

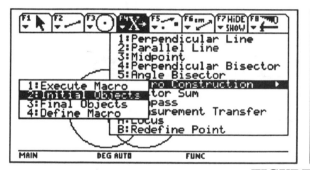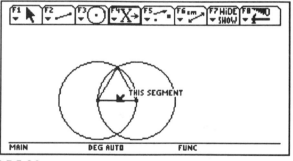

FIGURE M.32.
Macro construction step 1: choose the initial object. When you press ⌊ENTER⌋, the initial object becomes a dotted line.[38]

Now, go back to the Macro menu under ⌊F4⌋ and choose "3: Final Objects" , in the submenu of "6:Macro Construction". On the drawing, the desired final objects now must be highlighted.

Move the cursor to each of the two final segments in turn, and press ⌊ENTER⌋. As this is done, the segments will become dotted as they are in Figure M.33.

---

[38]If the macro required more than one initial object, you would move the cursor to each in turn, pressing ⌊ENTER⌋ after each object.

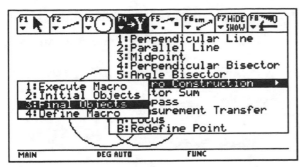 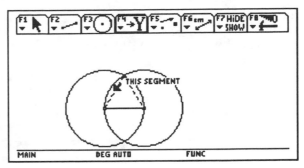

FIGURE M.33.

Macro construction step 2: choose the final objects (in this case, the segments which make up the second and third sides of the equilateral triangle).

Now, under ⬚, choose "6:Macro Construction" and then "4: Define Macro". A dialog box appears and an object name is required. Move the cursor to the second box and type a name such as *equitri*. Press ⬚ twice and another dialog box appears. This is the box that enables macros to be saved for use with other screens. If you don't put a name in here, the macro will only be defined for use with this screen. Since we do want to save it as a general macro, type the name *equitri* and press ⬚ twice. The keystrokes of this paragraph are exhibited in Figure M.34.

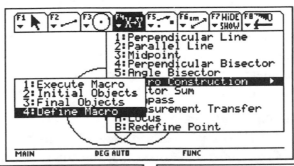

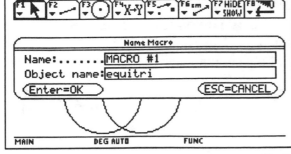 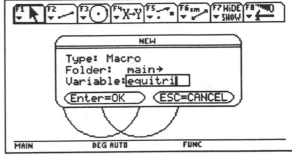

FIGURE M.34.

The final macro construction step: the macro is defined and named. The screen on the lower right is for saving the macro for use later.

Now, let's test the macro. First, clear the screen (select "8:Clear All" from the ⬚ menu. Now, construct a segment on the screen using the segment tool under ⬚. Be sure to construct this segment with the initial point to the left of the final point. After this is done, press [F4], choose "6:Macro Construction" and then "1:Execute Macro" on the submenu. See Figure M.35.

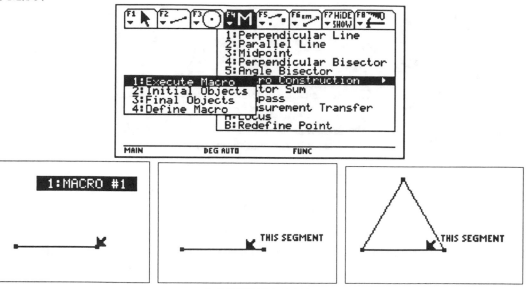

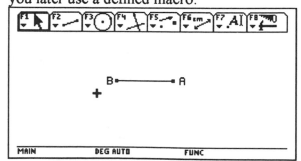

FIGURE M.35.

Press [ENTER] to begin executing the macro, move the cursor to the initial object (the segment), and press [ENTER]. An equilateral triangle immediately is constructed on the segment

Important: This macro used an initial segment that was drawn from left to right, and a final object, the equilateral triangle, was constructed *above* the segment. Now, see what happens when you construct a segment from right to left and then invoke the macro *equitri*. We labeled the segment below as it was drawn, A to B. (Remember you can label a point immediately after it is entered. In this case, to get a capital letter, use the shift key, [↑], and then the letter. The label for B was selected and dragged to its current location after typing it.) Now, execute the macro *equitri* and notice that the triangle has been drawn below the segment. Actually, the segment in the second example is rotated 180°, and hence the triangle that is constructed has been rotated by the same amount. (If you turn the calculator over so that the segment would appear to be constructed from left to right, the triangle is above the segment). Being aware of the direction of an initial object in a macro is very important when you later use a defined macro.

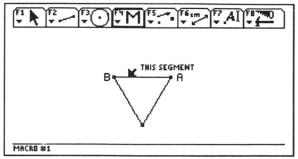

FIGURE M.36.
The order in which the initial segment is drawn affects the execution of the macro *equitri*.

M.5. [F5] *Transformation Menu*

The transformation menu contains tools for transformational geometry. Transformational geometry is a Euclidean Geometry which formalizes Euclid's tacitly assumed concept of superposition.

The [F5] tools for rotation and dilation differ from those on the [F1] menu, which operated in a freehand manner. With the [F5] tools, certain parameters such as angle, distance and direction must be specified.

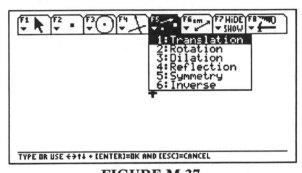

FIGURE M.37.
The [F5] Transformation Menu.

1: Translation

Let's begin with the translation tool. When an object is translated, it is moved a given distance in a given direction. Distance and direction define a vector that we must create in order to use this tool. To begin, let's create an object, such as a triangle upon which to work. First, clear the screen (choose "8:Clear All" from the [F8] menu). Then, select [F3], and create a triangle such as the one below by choosing "3:Triangle". Then, on the same screen create a vector using [F2], "7: Vector". See Figure M.38.

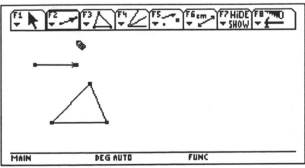

FIGURE M.38.
A triangle and a vector.

Now, select the tool "1: Translation" tool from the [F5] menu which is always done *before* selecting the object(s). Next, move the cursor to the triangle until the words "translate this triangle" appear. Press [ENTER] to select the triangle as in Figure M.39.

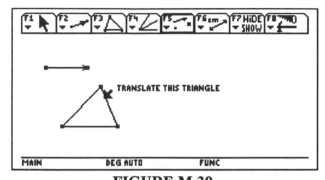

FIGURE M.39.
After choosing "1:Translate", select the triangle.

Then move the cursor to the vector until you see the words "by this vector". As soon as you press [ENTER], the triangle will be translated the required direction and distance. See Figure M.40.

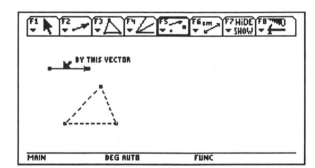 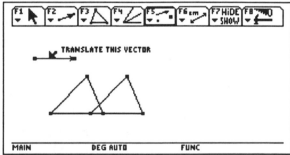

FIGURE M.40.
The triangle before (left) and after (right) translation by the vector.

At this point, both objects (before and after translation) are displayed. Note how the translation depends on the vector. Therefore, if the vector is changed, the triangle's new position will also change. Try it. Select the pointer tool, [F1], "1:Pointer". Then position the pointer at the head (arrow end) of the vector. Press [ENTER] to select and then hold down the hand key, ![hand], while moving the cursor in order to change the direction of the vector. Notice, the triangle is moving at the same time. When the desired direction and magnitude are achieved, press [ENTER]. See Figure M.41.

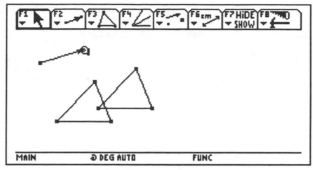

FIGURE M.41.

The translation was *relative* to the vector. So, as we change the vector (using the hand key), the translation of the triangle changes. (Compare with Figure M.40.)

2: Rotation

Before we use the F5, "2: Rotation" tool, recall that the pointer menu allowed freehand rotation about the geometric center of an object or about a defined point. Rotation through a specified angular value is also possible. However, we will need "Numerical Edit" from the F7 menu in order to specify an angle of rotation for this tool. We'll move ahead briefly to learn how to do this. Let's begin with a clear screen again: F8 "8:Clear All".

Now draw another triangle. Then press F7, " 6:Numerical Edit." After you press ENTER, move the cursor to any place on the screen you'd like to type a number and press ENTER. A box appears on the screen, with a text cursor inside. Type an angle measure, say 23. (You do not need to indicate units at this time. They are determined by the format settings and the default is degrees.) Then press ESC (not ENTER). See Figure M.42.

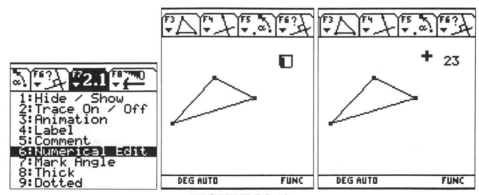

FIGURE M.42.

The numerical edit option is chosen *before* we rotate the triangle. Here, we enter 23°.

We're almost ready to rotate our triangle. Select F5 "2:Rotation" and then move the cursor to the triangle until the words "rotate this triangle" appear. Then, press ENTER. Since the triangle has been selected, it is now shown with dotted lines. Next, move the cursor to the number on the screen until the words "using this angle" appear and press ENTER. See Figure M.43.

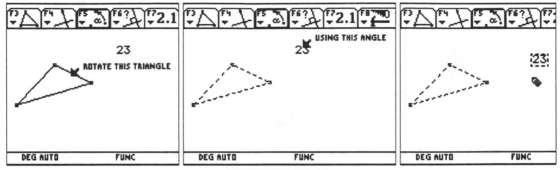

FIGURE M.43.
After choosing "2:Rotation", select the triangle and the angle measure.

To complete the operation, you will need to locate a point about which to rotate. As soon as you do that, the triangle is rotated. Move the cursor to a desired spot on the screen, such as the one pictured below, press ENTER and the triangle will be rotated. Again, notice that the original triangle still appears on the screen.

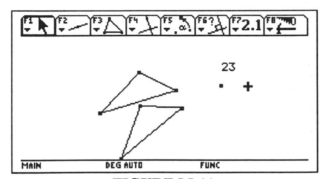

FIGURE M.44.
Select a point about which the triangle is to rotated and then press ENTER.

6: Inverse

Another interesting operation on the transformation menu is "6: Inverse." The inverse of a point M is always defined with respect to a given circle with radius r and center O. The inverse point N is located when $OM \cdot ON = r^2$ (that is, when the length of segment OM times the length of segment ON equals the square of the radius of the circle).

Clear the screen again (F8, "8:Clear All"). Now, construct a circle O (F3, "1:Circle") and a point M (F2, "1:Point"). The point M can be inside circle O, outside circle O or even on circle O. (If it is on the circle, the point and its inverse are one and the same.) In Figure M.45 below, M is inside the circle and was labeled immediately after entering the point.

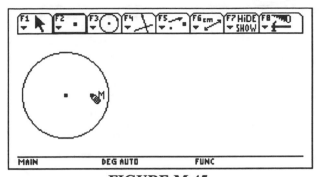

FIGURE M.45.
A point M is selected inside the drawn circle.

Now, press 🔲 ,"6:Inverse". Move the cursor to point M until the words "this point" appear. Then, press 🔲 . Point M will flash and have a dotted box appear around the label. Now move the cursor to the circle. The words "this circle" appear. As soon as you press 🔲 , the inverse point appears on the screen. In the screen on the far right in Figure M.46 below, the label N was typed immediately after the point appeared.

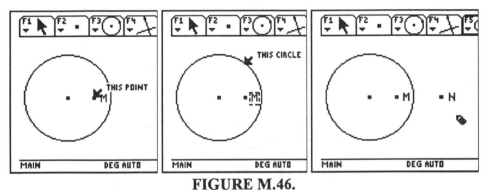

FIGURE M.46.
After choosing "6:Inverse", select the point (M), the circle, and then press 🔲 to have the inverse point (N) appear.

To get a feel for how the inverse of point M is affected by M's position and the circle's radius, move each around a little and observe what happens to the position of N.

An interesting thing happens when the locus of the original point M is on a circle that goes through the center of circle. Let's try it and see what happens. Clear the screen again to begin. Now, construct a circle, which is called the *circle of inversion*. Then, construct another circle that passes through the center. The second circle can be completely inside or cross to the outside of the original circle as long as it passes through the center of the circle of inversion. After selecting the center of the smaller circle, move the pencil to the center of the first circle. You will know this has occurred when you see the words "this radius point" appear on the screen as in Figure M.47.

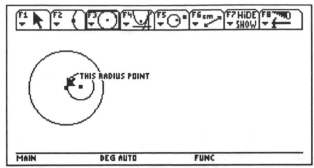

FIGURE M.47.

The second (smaller) circle passes through the center of the first circle.

Next, using F2, "2:Point on Object", move the cursor to the small circle and press ENTER to identify a point on the second circle. Now, find the inverse of the point you just placed with respect to the circle of inversion, using F5, "6:Inverse". First select the new point, then select the large circle. As soon as the second selection is made, the inverse point appears on the screen. See Figure M.48.

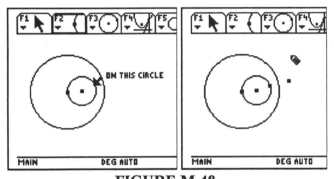

FIGURE M.48.

The inverse point is drawn on the right.

Using the pointer and the hand key, ![hand], drag the original point around the small circle and watch the movement of the inverse point. Can you predict its locus? Let's see if you are correct. Press F4, and select "A: Locus". Now, we need to indicate the point whose locus we want, and another point that lies on and moves around a path. Move the cursor to the constructed inverse point and press ENTER. The point begins to flash. Then move to the point on the inner circle. The screen on the right of Figure M.49 below was captured just prior to pressing ENTER after identifying the point on the smaller circle.

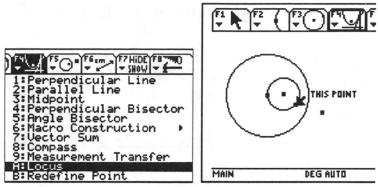

FIGURE M.49.
The first step in finding the locus of all inverse points.

As soon as you press [ENTER] you will see a straight line drawn as the locus of the inverse. It is depicted in Figure M.50.

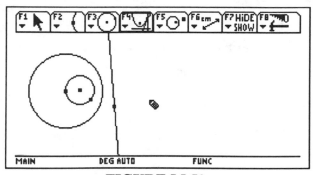

FIGURE M.50.
The straight line represents the locus of the inverse. Did you predict it?

Now, drag the point on the inner circle around the circle and watch the movement of the inverse point, up and down the straight line. As the point on the inner circle gets close to the center of the circle of inversion, the inverse (on the line) moves out toward an "ideal" point at infinity. (Remember the equation that defines the inverse.) This property--the locus of inverse points is a straight line when the path of the original point goes through the center of the circle of inversion--can be used, through linkages, to transform circular motion to straight line motion.

M.6.   [F6] *Measurement Menu*

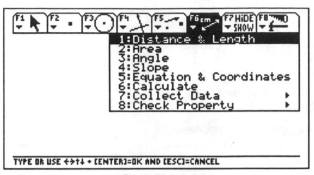

FIGURE M.51.
The [F6] Measurement menu.

Let's create a polygon as a practice object. Using [F2], "4:Polygon", create an irregular four-sided polygon by moving the cursor to each of the four desired vertices and pressing [ENTER] each time. Immediately after entering each vertex, type a letter to label the point (use the shift key, [↑] [A] etc.). After the polygon is complete, choose the pointer and move the labels to better positions by selecting each and using the hand key, 🖐. See Figure M.52.

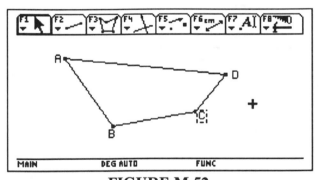

FIGURE M.52.
Construct and label an irregular four-sided polygon.

1: Distance and Length

To find the perimeter of this polygon press [F6] and select "1:Distance and Length." Move the cursor to one side of the polygon until the words "perimeter of this polygon" appear. Press [ENTER] and the perimeter (10.84 cm) appears next to the cursor. See Figure M.53.

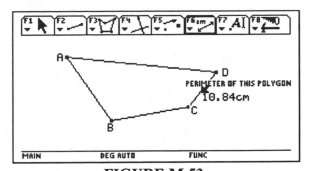

FIGURE M.53.

The perimeter is calculated using centimeters.[39]

To find the length of one side, the distance between two consecutive points is needed. Place the cursor on point A, press [ENTER]. A selection rectangle will appear around the label A. Then move the cursor to point B and press [ENTER]. The length AB is found to be 2.40 cm in Figure M.54 for our example. (Of course, the measurements for your polygon will vary.)

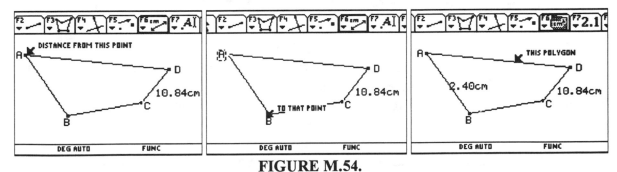

FIGURE M.54.

The length of segment AB is 2.40 cm.

Now, find the area. Press [F6] and choose "2:Area." Notice that the tool bar icon for [F6] has changed to an area icon. Move the cursor to a different side of the polygon for the sake of legibility. When the words "this polygon" appear, press [ENTER] and the area's numerical value (with appropriate units) appears on the screen. See Figure M.55.

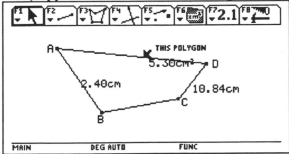

FIGURE M.55.

The area of the polygon is displayed as 5.30 cm^2.

---

[39]The type of unit is set under [F8], "9: Format."

3: Angle

Now, we'll measure an angle. Choose "3: Angle" from the [F6] menu. Then, move the cursor to the polygon in order to define the desired angle. Remember, it takes three points, with the vertex indicated by the center point. Mark the points on segment AD, vertex A and segment AB. As soon as you do this the angle measurement appears. See Figure M.56.

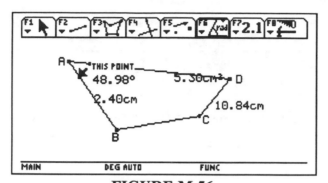

FIGURE M.56.
The measure of $\angle BAD$ is displayed as 48.98°.

6: Calculate

If a polygon is created by using segments, rather than the polygon tool, the figure is not recognized as a polygon, and the area or perimeter tools will not function. In such a case, to find these values, the calculate tool can be used. To begin, clear the screen by using the selection rectangle and pressing delete.

Let's consider a simple example of a triangle composed of three segments. That is, use the segment tool ([F2] "5:Segment")rather than the triangle tool ([F3]). When you draw it, be sure to make the endpoint of one segment the initial point of the next. Label each endpoint immediately after defining it. If you want to move a label, select it with the pointer tool and drag it to a new location using the hand key, [hand icon], and the cursor. Now, select [F6], "1:Distance and Length". Move the cursor to one side of the triangle. Notice the words "length of this segment" appear instead of "perimeter" in Figure M.57 then press [ENTER]. Do this for each side of the triangle.

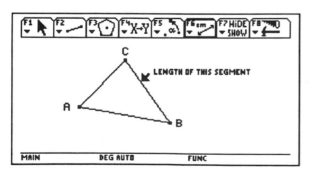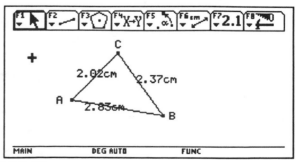

FIGURE M.57.
Since the triangle was constructed with the segment tool, selecting a side produces "length of segment" rather than "side of triangle". The lengths of the three sides are found on the right.

To find the perimeter now, press ⬚, and choose "6: Calculate." After you make this selection, a calculation box appears at the bottom of the screen. Also notice as you move the cursor up, that numerical values on the screen are highlighted sequentially. Since we want the perimeter, the lengths of the three sides will be added. Move the cursor until the length of BC is highlighted. Now, press ⬚. The side BC is labeled "a" and entered into the calculation box. See Figure M.58.

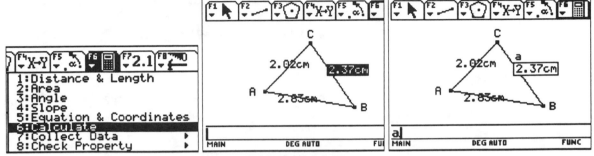

FIGURE M.58.
With the calculate tool, we begin to measure the perimeter by choosing the length of side BC.

Type ⬚ and then highlight another value by moving the cursor up. (If it moves to the right or left, its position in the calculation box is changed.) Enter the length of another side, press ⬚, then ⬚ and highlight and enter the length of the final side. Now press ⬚ and see the result (R:) displayed on the screen. You can drag this result (after selecting with the pointer) to any location on the screen. See Figure M.59.

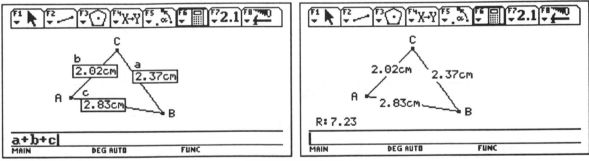

FIGURE M.59.
After each side has been selected, the result of the perimeter is displayed as R: 7.23.

8: Check Property

Another useful tool under ⬚ is "8: Check Property." First, let's construct two intersecting lines. Use the line tool under ⬚, "4:Line."

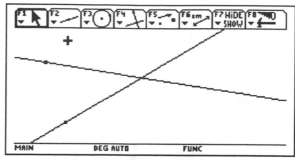

FIGURE M.60.

Begin by constructing two lines from the [F2] menu.

Then choose the "3: Perpendicular" property from the "8:Check Property" submenu of the [F6] menu. Move the cursor to one line until the words, "is this line" appear. Press [ENTER]. The line becomes dotted. Then move to the second line until the words "perpendicular to this line" appear. See Figure M.61.

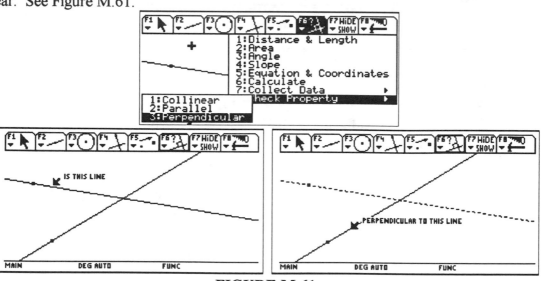

FIGURE M.61.

Checking the perpendicularity of two lines: first select the two lines.

When you press [ENTER] an empty box appears on the screen. You can move it to a new position simply by moving the cursor, or leave it where it is. In either case, you press [ENTER] again to see the result of the test as you can see in Figure M.62.

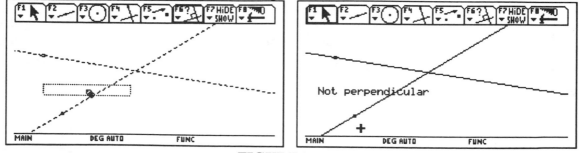

FIGURE M.62.

The result of the perpendicularity test: NO!

M.7. [F7] *Display Menu*

This menu allows you to choose tools which will change the way objects are displayed, as well as to trace and animate them. In addition, this menu gives you a tool to add and to edit numerical information, interactively.

```
1:Hide / Show
2:Trace On / Off
3:Animation
4:Label
5:Comment
6:Numerical Edit
7:Mark Angle
8:Thick
9:Dotted
```

Let's create an object on which to practice. First, clear the screen and draw a triangle such as the one below. Now, press [F7] and choose "4: Label." Move the cursor to a vertex until the words "this point" appear and press [ENTER]. An edit box with a cursor inside appears at the vertex. We'll label this vertex A. Therefore, type [↑] [A] and press [ESC]. See Figure M.64.

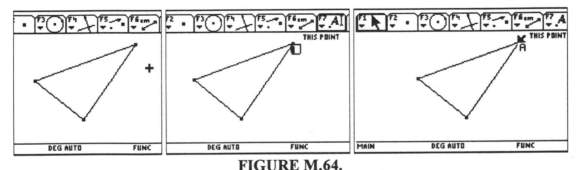

FIGURE M.64.
Constructing a triangle and labeling a vertex.

Repeat the process with vertex B and C respectively. You will have to reselect the label tool each time, since pressing [ESC] after each entry causes the selection to return to the pointer tool. Now, some of the labels should be repositioned. Select and drag as described earlier, using the pointer tool and hand key, 🖐. The letter will be shown in a box until the cursor is moved away and [ENTER] is pressed.

Now, let's mark an angle. Press [F7] and choose "7:Mark Angle." As always, three points are needed to define the angle. Let's mark angle C, so place a point on the side BC, at the vertex C and on the side AC. The points will flash on and off until [ENTER] is pressed. See Figure M.65.

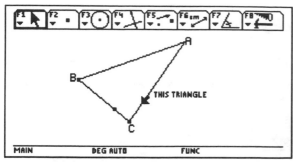 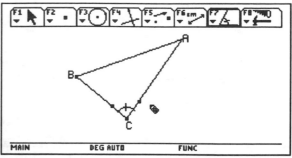

FIGURE M.65.

Left: Just prior to pressing ENTER for the third point. Right: As soon as you press ENTER, the angle mark is drawn.

(To define an exterior angle, drag the mark through the vertex to the exterior of the triangle.)

Now, let's measure the angle using F6, "3:Angle." Choose the three points that define your angle and press ENTER. The measure of the angle appears. (If, when you place your cursor near vertex C, the words "which object" appear, press ENTER. You will get a pop-up menu with your choices. You can choose either point, since they are the same. One is your original point C and the other is the point used to define the angle for marking.)

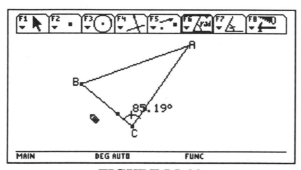

FIGURE M.66.

Finding the measure of $\angle ACB$.

Sometimes it is desirable to hide certain features or labels on a drawing. To illustrate the procedure, let's hide the measurement label. Press F7 and choose "1:Hide/Show." This works like a toggle switch on an object. Since, our object is now visible, move the cursor to the number until the words "this number" appear. Observe the effect in Figure M.67.

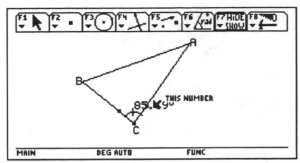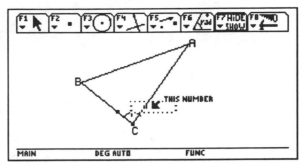

FIGURE M.67.
When you press ⟨ENTER⟩ the number is hidden and a dotted box appears on the screen.

This box will disappear when you choose another tool, or press ⟨ESC⟩ to go back to the pointer. Whenever the Hide/Show tool is on, dotted boxes and lines show the hidden objects. To see the object again, choose "1:Hide/Show" from ⟨F7⟩, move to the object and press ⟨ENTER⟩.

Similarly, we can change objects from solid to dotted, or from thin lines to thick. Choose ⟨F7⟩ and "9:Dotted." Move the cursor to the triangle and press ⟨ENTER⟩. In Figure M.68 you can see that the outline of the triangle changes from solid to dotted.

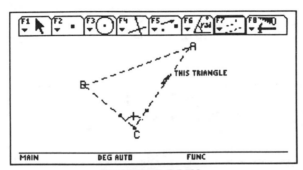

FIGURE M.68.
Changing the sides from solid to dotted.

6: Numerical Edit

Another useful feature of this menu is "6: Numerical Edit." We've already looked at this tool briefly but here we'll apply it to help construct a square. (Although the regular polygon tool would do this, frequently you need to construct a square on a given segment.) First, clear the screen. Now, draw a segment on the screen. Then press ⟨F7⟩ and choose "6:Numerical Edit." See Figure M.69.

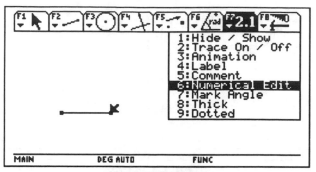

FIGURE M.69.

Selecting *numerical edit* for a given line segment.

Move the cursor to a convenient location and press [ENTER]. A box containing a text cursor inside appears on the screen. Type 90 (units will default to degrees) and press [ESC]. Now choose [F5], "2:Rotation." Remember, this is the rotation tool that allows you to specify an angle through which to rotate the segment. See Figure M.70.

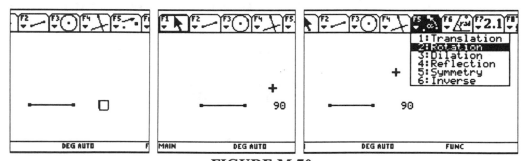

FIGURE M.70.

Preparing to rotate the line segment 90°.

Move the cursor to the segment and the words "rotate this segment" appear. Press [ENTER]. The selected segment becomes a dotted line. Now, move the cursor to the left endpoint of the segment. The words "around this point" appear. Press [ENTER] and move the cursor to the number 90, until you see the words "using this angle" as in the third screen in Figure M.71.

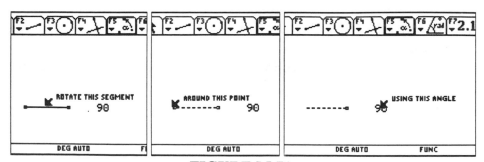

FIGURE M.71.

As soon as you press ENTER, another segment, rotated 90° from the first, is drawn. To complete the square, repeat the process twice, on the new segment and the next one created. The figures below show the square just prior to pressing the final ENTER and immediately afterward.

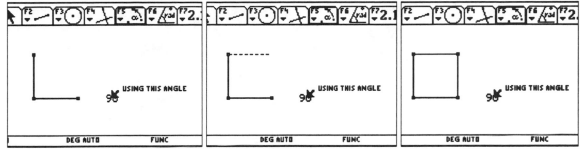

FIGURE M.72.

Left: completion of the first rotation. Center: constructing a third side. Right: the fourth side of the square.

NOTES

Exploration #1:
Linear Equations

Before you use the technology, you should understand that:

1. To solve a linear equation in one variable (say, x) we perform identical operations on both sides of the equation with the goal of reducing the original equation to a simple "$x =$" on one side of the equation. These operations are: simplification (that is, "removing parentheses"; addition of an appropriate quantity to both sides of the equation; and multiplication of both sides of the equation by an appropriate nonzero number.

2. A linear equation in two variables (say x and y) has many solutions, called *ordered pairs*, which correspond to points in the plane. These solutions collectively are represented by a straight line.

3. The one variable linear equation $3x - 2 = 10$ and the two variable linear equation $y = 3x - 2$ have the common (x) solution of 4 when $y = 10$.

4. The one variable linear equation in x, $ax + b = 0$, has the same x-value solution as the x-intercept of the equation $y = ax + b$.

Procedures

Procedure #1. Let's begin by entering the following equation into the *TI-92*: $3x - 2 = 10$. The strategy for solving this equation for x is to first add two to each side of the equation to obtain $3x = 12$. Next, we divide by both sides by 3 (or equivalently, multiply both sides by $\frac{1}{3}$). One way to perform these steps on the *TI-92* is depicted in Figure 1.1 below.

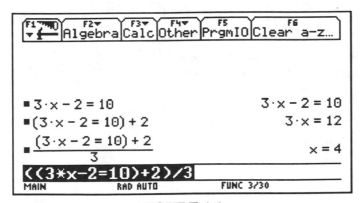

FIGURE 1.1.

Procedure #2. Now let's examine the related two-variable equation, $y = 3x - 2$. On the *TI-92*, we enter $3x - 2$ for $y1$ in the $\boxed{\text{Y} =}$ editor. For the $\boxed{\text{WINDOW}}$ values, choose xmin $= -2$, xmax $= 9.9$, xscl $= 2$, ymin $= -4$, ymax $= 16.4$, and yscl $= 2$. Since we are interested in the value of x when $3x - 2$ equals 10, we press $\boxed{\diamond}\boxed{\text{GRAPH}}$ and, using $\boxed{\text{F3}}$ Trace, press the left and right cursor keys until the y-coordinate ("yc:") is equal to 10. In Figure 1.2 we see that the corresponding x value is (the solution) $x = 4$.

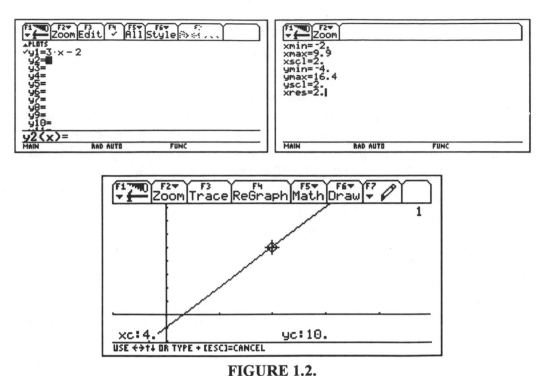

FIGURE 1.2.

With the trace (F3) feature on, we see that $3x - 2$ equals 10 when x equals 4.

Procedure #3. The one-variable equation $3x - 2 = 10$ is equivalent to the *standard-form* equation $3x - 12 = 0$. We can see that by subtracting 10 from both sides of $3x - 2 = 10$, we obtain a zero on the right side to get $3x - 12 = 0$. Now, it is important to know that solving $3x - 12 = 0$ is equivalent to finding the x-intercept of the two-variable equation $y = 3x - 12$.[1] The "Zero" option of the F5 Math key in the graph window is another way to find that x-intercept. The *TI-92* prompts you for a lower and an upper limit for that zero. See Figure 1.3 and 1.4 for the screen displays that gives us yet a third confirmation that indeed, the solution to $3x - 2 = 10$ is the x-coordinate of $y = 3x - 12$, $x = 4$!

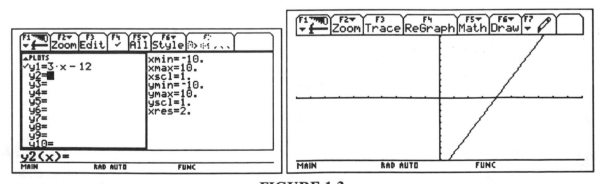

FIGURE 1.3

The graph of $y = 3x - 12$ with the ZoomStd viewing window.

---

[1]An x-intercept is simply the x-coordinate of a point on a graph (if any) which lies on the x-axis. The y-value of that point is zero.

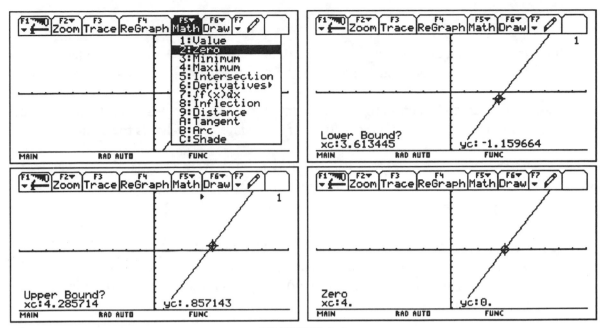

FIGURE 1.4

Upper left: choose "2: Zero" from the [F5] Math menu. Upper right: move cursor just left of intercept; press [ENTER].
Lower left: move cursor just right of the x- intercept; press [ENTER]. Lower right: *TI-92* displays the zero as "xc:4".

Exercises

In questions 1 through 6, use Procedure #1 as a guide to solve each equation for the variable x.

1. $3x + 2 = 8$ 2. $2x + 7 = x + 4$ 3. $4x - 3 = 1 + 5x$

4. $2(x + 1) = 1 - x$ 5. $2x - 5 = -1 - 4x$ 6. $3(x - 2) = 5x + 9$

In questions 7 through 10, use Procedure #2 as a guide to solve each equation graphically. Use the same window values as are displayed in Figure 1.2.

7. $2x + 1 = 7$ 8. $-2(3 - x) = 9$

9. $-\frac{1}{2}(2x - 1) = -3$ 10. $2(-x + 3) = 8$

Use the *TI-92*'s "Zero" finding capability (as outlined in Procedure #3) to graphically solve each of the equations in questions 11 through 14.

11. $4x + 1 = 7$ 12. $-2(3 - 2x) = 9$

13. $-\frac{1}{2}(4x - 1) = -3$ 14. $2(-4x + 3) = 8$

15   There is not a unique solution to the equation $-3(2 - x) = 3x - 6$. What happens when you try to solve this equation using the method of Procedure #1? What is the solution?

16. There is not a unique solution to the equation $2(3x - 4) = 6x + 1$. What happens when you try to solve this equation using the method of Procedure #1? What is the solution?

17. The relationship between degrees Fahrenheit (F) and degrees Celsius (C) is given by the formula $F = \frac{9}{5}C + 32$. There is one temperature that is the same value on both scales. What is that temperature? (Hint: Solve the equation $\frac{9}{5}C + 32 = C$.)

18. The formula for the perimeter P of a rectangle of length l and width w is $P = 2l + 2w$.
   a) Use the method of Procedure #1 to solve this equation for w.
   b) Enter the command solve(P=2l+2w,w) and compare the *TI-92*'s response to your answer in part a.

Exploration #2:
Equations of Lines

Before you use the technology, you should understand that:

1. The slope of a line is a measure of the line's inclination (or angle) with respect to the x-axis. If (x_1, y_1) is one point on the line and (x_2, y_2) is a different point, then the slope m is given by

$$m = \frac{y_2 - y_1}{x_2 - x_1}$$

2. The equation of a non-vertical line passing through (x_1, y_1) and (x_2, y_2) is:

$$y = m(x - x_1) + y_1 \quad \text{or} \quad y = \frac{y_2 - y_1}{x_2 - x_1}(x - x_1) + y_1$$

3. The *slope/intercept form* of a straight line is $y = mx + b$ where b is the y-intercept and m is the slope.

4. The *standard form* of a straight line is $Ax + By = C$ which, when rewritten as $y = -\frac{A}{B}x + \frac{C}{B}$ can be seen to have slope $-\frac{A}{B}$ and y-intercept $\frac{C}{B}$.

5. Vertical lines have the form $x = c$; horizontal lines have the form $y = c$ where c is some real number.

Procedures

Procedure #1. It is very useful to be able to determine the equation of a line given two of its points (or its slope and one of its points). Let's begin by defining a *TI-92* function called lineq for finding the equation of a line given two points (a_1, b_1) and (a_2, b_2):[2]

$$\text{lineq(a1,b1,a2,b2)}=((b2 - b1)/(a2 - a1))*(x - a1) + b1$$

For example, suppose we want to find the equation of a line which goes through the points $(1, 5)$ and $(-2, -1)$. We simply have to enter lineq(1,5,-2,-1) and the *TI-92* will display the right side of the line's equation, $y = 2x + 3$. See Figure 2.1.

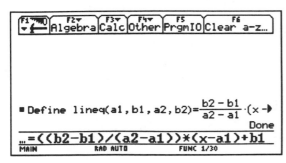 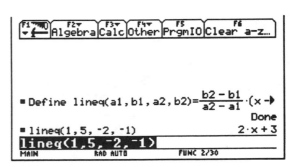

FIGURE 2.1

Left: Defining the two-point equation.

Right: Invoking the new function to find the equation of the line passing through $(1, 5)$ and $(-2, -1)$.

---

[2]We cannot use (x_1, y_1) and (x_2, y_2) here because y_1 and y_2 are *reserved variable names* for the *TI-92*. They are two of the names of the list of functions in the $Y =$ screen.

The easiest way to graph the equation $y = 2x + 3$ now is to highlight $2x + 3$ and copy it to the buffer with ◇ C. Then, enter the Y= editor and paste $2x + 3$ in for $y1$ with ◇ V. In Figure 2.2, we depict the graph after turning "Grid" on from the F1 9:Format submenu on the GRAPH screen.

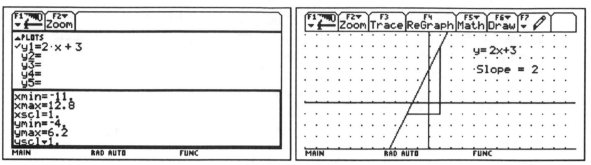

FIGURE 2.2.
The graph is labeled using the "Text" from the F7 menu. (We also drew the vertical and horizontal line segments from that menu.)

Procedure #2. The *TI-92* has a built-in function for drawing the graph of a straight line given a point on it and its slope. For example, if we want to see a graph of the line that passes through the point $(2, 3)$ with slope -2, we can enter the following on the edit line in the HOME screen editor:[3]

$$\text{DrawSlp } 2, 3, \ -2$$

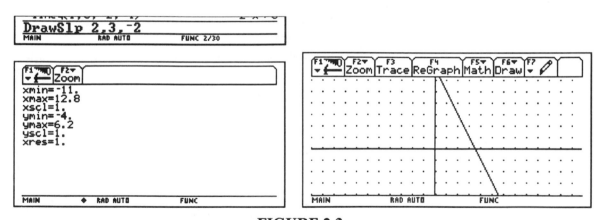

FIGURE 2.3.
The line with slope -2 passes through the point $(2, 3)$.

The *equation* of the line passing through $(2, 3)$ with slope -2 can be found by using the point-slope form of a straight line: $y = m(x - x_1) + y_1$. If we had to find several such

_____

[3]To get the graph in Figure 2.3, be sure to first clear any entries in the $Y =$ editor. Also, the DrawSlp function can be entered directly from the keyboard (pay close attention to the capital letters D and S) or it can be obtained from either the catalog or the F6 Draw menu in the Graph editor.

equations, we might want to enter this formula as a function just as we did with the function *lineq* in Procedure #1. Using the function name *ptslope*, we do just that.[4] See Figure 2.4.

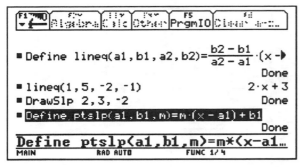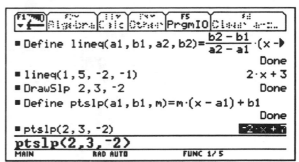

FIGURE 2.4.

The equation of the line passing through $(2, 3)$ with slope $m = -2$ is $y = -2x + 7$.

Exercises

In questions 1 through 6, a) use the *lineq* function discussed in Procedure #1 as a guide to find the equation of the straight line passing through each pair of points; b) graph the line.

1. $(1, 3)$ and $(2, 5)$ 2. $(3, 7)$ and $(-2, -3)$

3. $(1, 3)$ and $(2, 7)$ 4. $(-1, -1)$ and $(-3, 3)$

5. $(0, 0)$ and $(-\frac{1}{2}, 1)$ 6. $(1, 2)$ and $(-1, -4)$

7. What happens to the function *lineq* when you try to use it to find the equation that passes through the points $(1, 5)$ and $(1, 4)$? Can you explain this behavior?

8. Create a *TI-92* function that finds the slope connecting two points. Call it SLP1.

9. Redefine the *lineq* function to incorporate the function you created from question 8.

10. Consumer spending is something watched closely by retailers. This is especially true for the period between Thanksgiving and Christmas, the heaviest shopping period of the year. For ten years (1985 to 1994), Americans' spending for the holiday period is given below:

---

[4]Notice the use of the asterisk for multiplication in "$m*(x\text{-}a1)$..." It is necessary because implied multiplication will not work in cases like this.

Year	Amount Spent (in Billions of $)
1985	$28.0
1986	31.4
1987	37.0
1988	42.8
1989	48.9
1990	56.2
1991	59.8
1992	66.8
1993	79.1
1994	96.9

A reasonable approximation to this spending (which is surprisingly nearly linear) is given by:

$$y = 7.038x - 13947.8$$

a) Find y for $x = 1998$. What can you conjecture, using this model, about consumer spending between Thanksgiving and Christmas in 1998?

b) Find y for $x = 1987$ and compare this number with the number in the table. You can see in the plot below that 1987 spending falls very close to the line. This is true for 1993 as well.

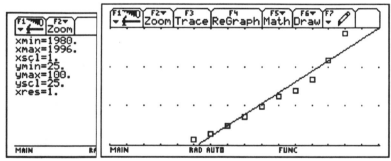

Exploration #3:
Concurrency of the Altitudes of a Triangle

Before you use the technology, you should understand that:

1. Determining whether several lines are concurrent can be very challenging. There are some theorems that are helpful, but sometimes looking at the lines in a different context can be helpful. In this exploration the altitudes of a triangle will be created and a sketch of a possible proof of concurrency will be suggested. You should be familiar with the definition of the perpendicular bisector of a line.

2. Perpendicular bisectors of the sides of a triangle are concurrent.

Procedures

Procedure #1. To begin, we construct a scalene triangle as in Figure 3.1.

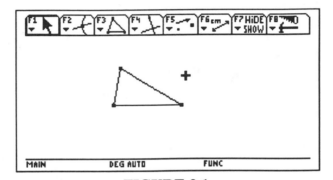

FIGURE 3.1.
First step: construct a scalene triangle.

Now, use the perpendicular line tool (press [F4] then choose "1:Perpendicular Line")three times to construct perpendicular lines to each side from the opposite vertex. These lines are called the *altitudes* of the triangle. See Figure 3.2.

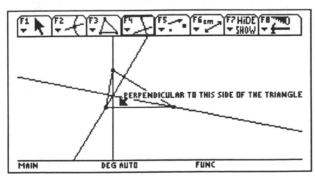

FIGURE 3.2.
The three perpendicular lines passing through opposite vertices are constructed.

Do the lines appear to have any special properties? It should appear that these lines are concurrent. Is this just a coincidence? Change the shape of the triangle by grabbing a vertex and pulling it. What seems to be happening to those altitudes? Vary the triangle to a variety of different shapes. It should appear that the altitudes remain concurrent for any triangle. A proof is outlined in this exploration.

Using the parallel line tool (press F4 then choose "2:Parallel Line") construct lines parallel to each side of triangle through opposite vertices as in Figure 3.3.

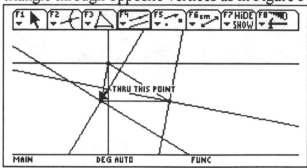

FIGURE 3.3.

Lines are drawn parallel to each side of the original scalene triangle, through opposite vertices.

Now, define the points of intersections of the new lines so that they can be labeled. First, choose F2 "3:Intersection Point". Then move the cursor to one pair of the intersecting lines and press ENTER. Do the same for the other pairs of intersecting lines you have constructed. Now, label the points of intersection of these new lines. To do this, under F7 choose "4:Label". Then move the cursor to the desired point of intersection (the words "this point" appear) and press ENTER. A text box appears and enter the label "A" and proceed to label the remaining points as "B" and "C". If the labels need to be moved to a different position, select them with the pointer (F1 , "1:Pointer") and drag them to a new location. The label for C is being moved in Figure 3.4 below. Notice, these new lines form another triangle. Change the altitude lines to dotted lines using F7 "9:Dotted" to make the picture a little less cluttered.

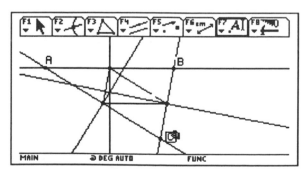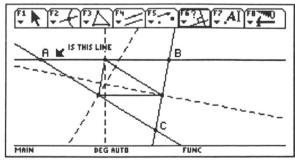

FIGURE 3.4.

Left: the three points are labeled (C's label is in the process of being moved). Right: the altitude lines are dotted to cut down on clutter.

Procedure #2. It looks like the altitudes of the inner triangle play another role in determining the large triangle. In fact, you may want to investigate any special properties you see. Here, we concentrate on the fact that it appears as if *the altitudes of the inner triangle are perpendicular to the sides of the large triangle*. That property can be checked by using the tool under [F6], "8: Check Property". Select "3:Perpendicular" now and move the cursor to one of the desired lines until the words "is this line" appear (as it does on the right side of Figure 3.4). Press [ENTER]. Then move the cursor to the line for which you are checking perpendicularity. The words "perpendicular to this line" appear. See Figure 3.5.

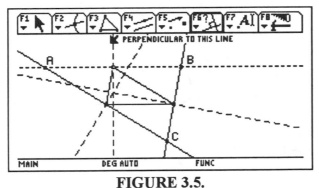

FIGURE 3.5.
Checking perpendicularity.

Press [ENTER], and a box appears on the screen. When you press [ENTER] again, you will see the conclusion of this test. (Prior to pressing [ENTER], the box can be moved to any desired location by moving the cursor.) The results of this test appear in Figure 3.6.

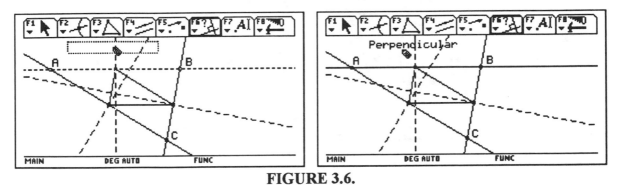

FIGURE 3.6.
Left: The box appears as the test starts. Right: The conclusion is that the lines are, in fact, perpendicular.

Repeat the process for the other intersections of the altitudes of the inner triangle with the sides of the large triangle. Notice, in all cases the tested lines are perpendicular.

If the diagram is studied again, it should appear that not only are the lines perpendicular, but the altitudes of the small triangle might bisect the sides of the large one. To check that, use the measurement tool and measure the length of each side from the intersection point to the vertex. This is accomplished by choosing [F6] "1:Distance & Length" and then moving the cursor to vertex A. You will see the words "distance from this point" appear. Press the

[ENTER] key. Now, move the cursor to the intersection of the altitude with side AB. The words "to that point" appear. When [ENTER] is pressed, the length of that segment appears. In Figure 3.7 that length is seen to be 1.97 cm.

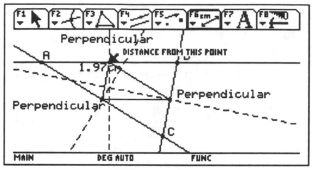

FIGURE 3.7.
The distance from A to the intersection of AB with the altitude is 1.97 cm.

Repeat the process and measure the segment from that point of intersection to vertex B. The two segments are the same length (1.97 cm) and are perpendicular to the altitude of the small triangle. Now, check the measurements of the segments on the other two sides of triangle ABC. In Figure 3.8 below all measurements and properties appear. The labels have been moved for clarity.

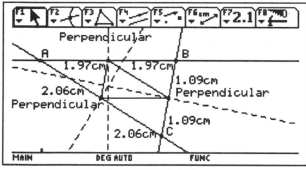

FIGURE 3.8.
Conclusion: The altitudes of the small triangle are the perpendicular bisectors of the large triangle!

We have seen that the altitudes of the small triangle are the perpendicular bisectors of the large triangle. Since the perpendicular bisectors of a triangle are concurrent, the altitudes of the small triangle are also concurrent, because these are the same lines. The small triangle had no special properties, so, *in general, the altitudes of a triangle are concurrent*.

Exercises

1. Do you see any other properties of the triangles you constructed? For example, the large triangle is divided into four small triangles; how are they related? Do you see any parallelograms? Can you prove they are parallelograms?

2. It is important, in these geometry explorations, to be sure to duplicate the screens we display. Then, try grabbing an early point and observe if certain properties still hold. For example, for the screen shown in Figure 3.3, grab one of the vertices of the original scalene triangle and observe what happens to the altitudes.

3. Write, in your own words, what properties of triangles you observed to be true from this exploration in geometry.

4. You have seen that the altitudes of a triangle are concurrent. This point is called the *orthocenter* of the triangle. Where is the orthocenter located when the triangle is a(n):
   a) acute triangle
   b) obtuse triangle
   c) right triangle

   Construct an example to illustrate each case.

5. A very simple proof of the concurrency of lines is possible using **Ceva's Theorem.** This was developed in 1678 by the Italian mathematician Giovanni Ceva. Investigate this theorem and use it to prove that the altitudes of a triangle are concurrent.

NOTES

Exploration #4:
Locating a Minimal Distance Point

Before you use the technology, you should understand that:

1. The definitions of equilateral triangle and angle bisector.

2. The sum of the angles of a triangle is 180°.

3. The sum of the angles surrounding a point is 360°.

4. The measure of a central angle is equal to the intercepted arc.

5. The measure of an inscribed angle is $\frac{1}{2}$ of its intercepted arc.

Procedures

Procedure #1. Finding a point, which when connected to the vertices of a polygon, gives a path of minimal distance is a useful model of some real world problems. Whether you are trying to put in a sidewalk which connects your front door, garage and mailbox and uses the least amount of concrete, or whether you are trying to locate a point which will connect a number of telephone lines using the least amount of cable, you would like to find a solution that will use less material than simply traveling around the edges of the polygon formed by connecting these items. The minimal distance point can be located physically using a plexiglass model and soap film (i.e. soap bubbles). When this is done for triangles, the solutions exhibit a common property which is that all paths meet at exactly 120° angles.

This exploration locates a point P in a triangle such that the sum of the lengths of the segments from P to each vertex will be a minimum. Equivalently, we say that the line segments form a *minimal path*. The problem can be restated:

Problem: Given △ABC, with no angle greater than 120°, locate point P so that ∠APB ≈ ∠APC ≈ ∠BPC = 120°.

First, construct △ABC, labeling each point as it is created. *The order in which the sides are drawn is very important, because the predefined macro which draws equilateral triangles will be used.* See Figure 4.1.

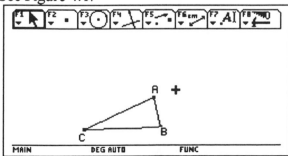

FIGURE 4.1.
We begin with a triangle having every angle less than 120°.

Now, we open the macro called *equitri* by pressing  selecting "1:Open"[5].   Choose "2:Macro" and enter its name, *equitri*. (You will have to press ENTER twice.)   See Figure 4.2.

FIGURE 4.2.

The macro called *equitri* is opened for this exploration.

Now, press F4 and select "6: Macro Construction" followed by "1: Execute Macro." To actually invoke the macro, press ENTER when the macro name appears in reverse video. (See the screen on the upper left of Figure 4.3.)   Next, we move the cursor to side AC of the triangle and press ENTER. You immediately see an equilateral triangle built on side AC. (Notice, if the triangle had been constructed in a counterclockwise direction, the new equilateral triangle would be on the inside of our triangle, rather than the outside.)

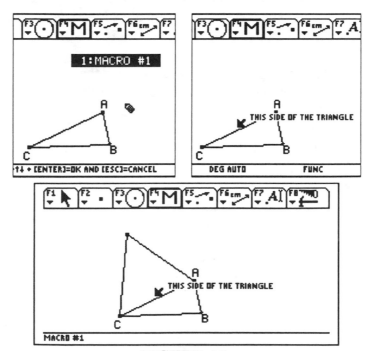

FIGURE 4.3.

By invoking the macro *equitri* on side AC, an equilateral triangle is drawn.

---

[5]This macro was created in the *Overview*. If you did not create and save that macro already, you must do it before continuing with this exploration. See pages 80-84.

Now, we want to repeat the macro drawing of an equilateral triangle on side AB. So, move the cursor to the side AB and press ENTER. Now that two new triangles have been created, label their vertices D and E respectively. See Figure 4.4.

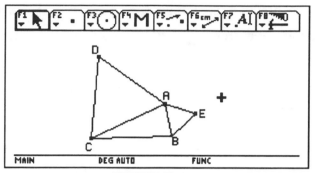

FIGURE 4.4.
Triangles ACD and ABE are equilateral triangles constructed by the macro *equitri*.

Keep in mind that all the angles of the new triangles ACD and ABE equal 60° since they are equilateral triangles. Now, we want to bisect angles BAE and ABE. Locate the intersection point of the bisectors and label it F as we have done in Figure 4.5.

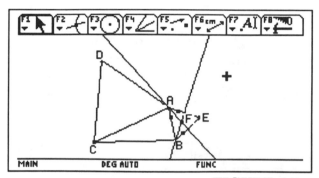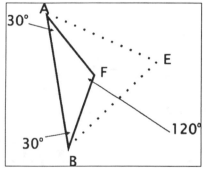

FIGURE 4.5.
Left: the bisectors of ∠BAE and ∠ABE meet at point F. The interior angles of △FAB are labeled on the right.

Note that angles FAB and FBA are each 30° since they are half of a 60° angle. Therefore angle AFB is 120° since the sum of the angles of a triangle is 180°. Now, we construct a circle with the center at F and radius FB (or FA). See Figure 4.6.

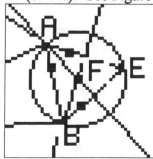

FIGURE 4.6.
A circle of radius FB is drawn with center at point F.

Notice that any point on arc AB will form a 120° angle when connected to points A and B since it intercepts a 240° arc. (Remember the central angle is 120°.) Now, to find the unique point needed on arc AB, we will repeat the construction on the other equilateral triangle. But first, to help keep the drawing from appearing too cluttered, we hide the construction lines for the previous angle bisectors. Choose F7, "1: Hide/Show" and then move the cursor to each line and press ENTER. The lines will appear dotted as below in Figure 4.7, until another tool is selected.

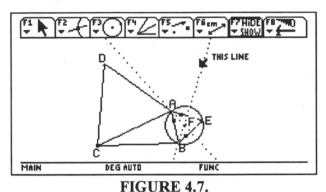

FIGURE 4.7.
The angle bisectors have been hidden.

Now, consider the other equilateral triangle, △ACD. To repeat the construction of angle bisectors and circle on △ACD, we bisect angles DCA and DAC. Then, we locate the intersection point and label it G. The circle is constructed at G and has radius GC. You will see that the two circles intersect at vertex A and at a second point. This second point is labeled as point P in Figure 4.8.

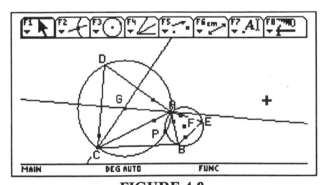

FIGURE 4.8.
The two circumscribed circles meet in two points, A and P.

Procedure #2. *Point P is the desired point of minimal distance.* If it is connected with the vertices of the original triangle, it will make 120° angles with each vertex. Now, we concentrate on the original triangle and the newly found point P so we hide the other construction lines. Using the segment tool (F2 "5:Segment"), we draw the segments that connect P to each vertex. These segments form the minimal path. If A represented the front door, B the garage and C the mailbox, this is the path that should be paved in order to use the least amount of cement.

By just using [F7] "Hide/Show" to hide most of the construction, in Figure 4.9 we show the original triangle with the path on the left and the path only on the right.

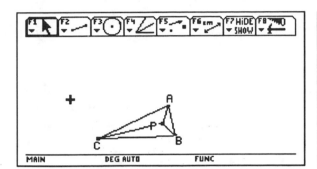 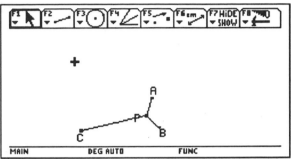

FIGURE 4.9.
The segments are drawn. P has the property that PA + PB + PC is minimal.

We conclude this exploration by mentioning that the point P we have located connects to three given points and yields a minimal path connecting them. Point P is called a **Steiner point** and the path illustrates a **Steiner tree.**

Exercises

1. Does the point P correspond to any other common point in your triangle? For example, is it the circumcenter, incenter centroid or orthocenter.?

2. A Steiner point exists for all triangles that have angles less than 120°. Where is it located if one of the angles is exactly 120°? (There is no Steiner point if the triangle has an angle greater than 120°. The shortest distance would be found by simply connecting the three points with two line segments.)

In questions 3 through 8, you may find the final figure in this exploration helpful in forming your answers.

3. The three point problem gives some general information about Steiner trees. Since in a Steiner tree, every angle measures exactly 120°, what is the maximum number of edges that connects every given point to the tree?

4. How many edges exist at every Steiner point?

5. How are the number of edges in the tree related to the number of given points plus Steiner points?

6. Since exactly three edges meet at every Steiner point and at least one edge must touch every point, what is the maximum number of Steiner points in any problem? (How is it related to the number of given points?)

7. How many Steiner points would there be for a square? Can you find them? (Reduce your problem to that of the triangle.)

8. Steiner points can be found for other polygons. If you form a rectangle, you will find two pairs of points, one set parallel to the long side and one set parallel to the short side. Can you locate these? These both give "relative" minimum solutions. Which set gives the shortest path?

Exploration #5:
Napoleon's Theorem

Before you use the technology, you should understand that:
1. The angle bisectors of the angles of an equilateral triangle are concurrent.

2. The perpendicular bisectors of the sides of an equilateral triangle are concurrent.

3. The altitudes drawn to each side of an equilateral triangle are concurrent.

Procedures
Procedure #1. Napoleon Bonaparte, the general and emperor of France, had a considerable interest in geometry and is credited with revolutionizing the teaching of mathematics in France. Several mathematics historians credit his reforms for the great increase in creative mathematics in 19th century France.

He is credited with discovering the following theorem:

> **Napoleon's Theorem:** *Given any triangle, construct equilateral triangles on its sides. Then connect the* centroid *(or* circumcenter, *or* incenter) *of each of the new triangles to one other, forming a new triangle. This new triangle will always be equilateral.*

Let's go through the construction.

First, draw any scalene triangle using segments. We use segments to enable us to use the macro, *equitri*, that was defined earlier. We want the new triangles to be on the outside of the given triangle. Therefore, remember to be careful about the direction in which you draw the segments. Our macro, *equitri*, operated on a segment drawn *from left to right* and the triangle was constructed *above* that segment. Thus, after you draw the first segment, continue the next segments in a clockwise sense. This assures that the Macro will construct only external equilateral triangles. The scalene triangle is drawn in Figure 5.1.

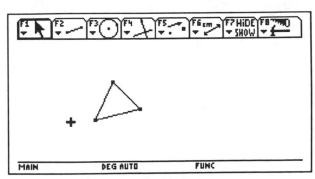

FIGURE 5.1.

We begin with a scalene triangle using [F2] "5:segment" and constructing in a clockwise sense.

Now, we open the macro called *equitri* by pressing [F8] selecting "1: Open"[6]. Choose "2:Macro" and enter its name, *equitri*. (You may have to press [ENTER] twice.) See Figure 5.2.

FIGURE 5.2.
The macro called *equitri* is opened for this exploration.

Now, press [F4] and select "6: Macro Construction" followed by "1: Execute Macro." To actually invoke the macro, press [ENTER] when the macro name appears in reverse video. (See the screen on the left of Figure 5.3.) Now select each segment and in turn press [ENTER] after selecting each. An equilateral triangle will be drawn on each segment as we see on the screen on the right in Figure 5.3

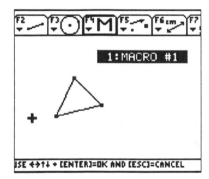 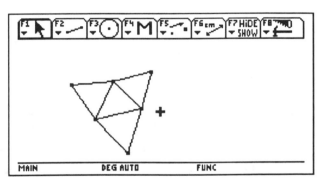

FIGURE 5.3.
By invoking the macro *equitri* on each side, three "external" equilateral triangles are drawn.

Since our new triangles are equilateral, the angle bisectors, perpendicular bisectors of the sides and altitudes all intersect in the same point (are concurrent) for each triangle. You can use any of these to locate that point. Since the *TI-92* has a perpendicular bisector tool, that will be the easiest way. In addition, since the perpendicular bisectors are concurrent, we need only to construct two in each triangle to locate the desired point. Press [F4], "4: Perpendicular Bisector." Then, move the cursor to one side of one of the new triangles, press [ENTER] and repeat on another side. The perpendicular bisectors will be drawn, and Figure 5.4 shows the first two.

---

[6]This macro was created in the *Overview*. If you did not create and save that macro already, you must do it before continuing with this excursion. See pages 80-84.

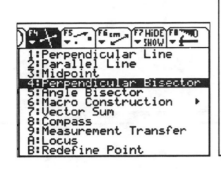
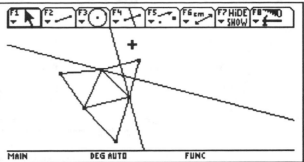

FIGURE 5.4.

Two perpendicular bisectors are constructed on one of the equilateral triangles.

Now, we really just want the point of intersection of those bisectors, so let's define it. Press [F2], "3: Intersection Point". Then move the cursor to one of the bisectors until you see the words "this line" and press [ENTER] . The line becomes dotted. Now repeat the procedure with the other bisector. As soon as you press [ENTER], the point at the intersection of the two lines will be drawn.

Since we are only interested in the *point* of intersection, let's hide the extra lines which clutter the drawing. Press [F7], "1: Hide/Show." Then move the cursor to each of the perpendicular bisectors, and press [ENTER]. The lines become dotted, and then disappear when you change tools. Finally, press [ESC] and they will disappear from view, leaving only the point of intersection. See Figure 5.5.

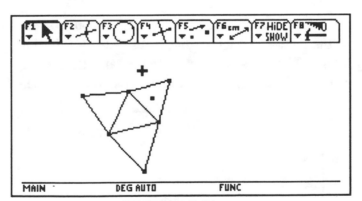

FIGURE 5.5.

The original triangle, the macro-constructed equilateral triangles, and the point of intersection of the perpendicular bisectors in one of those equilateral triangles.

We need to now repeat this procedure (of finding the "center" of each equilateral triangle) for the other two equilateral triangles, leaving only the points of intersection of the perpendicular bisectors. The figures below show the hidden lines before pressing [ESC] and immediately after. See Figure 5.6.

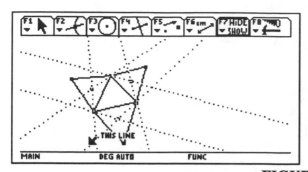
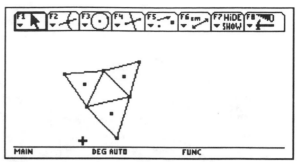

FIGURE 5.6.
All three intersection points have been found!

Recall what we need to show: that these three points form the vertices of an equilateral triangle. So, we will connect the three points we have just constructed using the segment tool. See Figure 5.7.

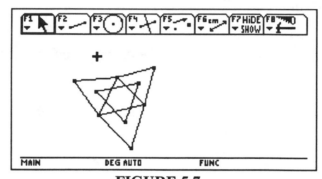

FIGURE 5.7.
All three points from Figure 5.6 are connected using the segment tool.

Let's measure the sides of the triangle we have just constructed. Press $\boxed{\text{F6}}$, "1: Distance and Length." Move the cursor to each segment, respectively. In the figure below, the three lengths have been found. The sides are seen to be the same length in Figure 5.8. (Some of the numbers have been dragged to a slightly different position); hence, ***the triangle is indeed equilateral!***

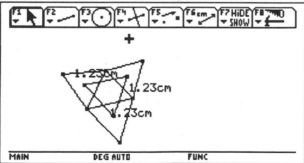

FIGURE 5.8.
The three points from Figure 5.7 do in fact form an equilateral triangle.

Procedure #2. Is this just some coincidence related to the shape of our original triangle? Let's see. Drag a vertex of the original triangle to a new position. (Be patient. It moves very slowly.) Watch how the measurements change.   What you observe is one of the more powerful features of the *TI-92*: that all the steps we performed in the Geometry screen are *relative*. See Figure 5.9.

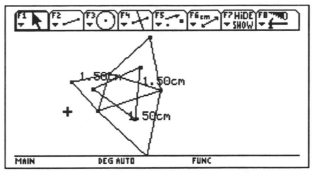

FIGURE 5.9.
Even as the original triangle is changed, the three circumcenter points form an equilateral triangle.

Conclusion: No matter how you stretch or shrink your original triangle, the triangle formed by connecting those circumcenters remains equilateral!

Exercises

1. Connect each vertex of the original triangle to the farthest vertex of the opposite equilateral triangle.
    a) Do the three lines appear to be congruent?
    b) Can you prove that they are congruent?

2. The lines you drew above appear to be concurrent. Can you prove they are? You will need to work with the circumscribed circles of the equilateral triangles.)

3. In our construction, we used the perpendicular bisectors of the sides of the equilateral triangles. Why would this also work with the angle bisectors or altitudes

NOTES

Exploration #6:
Polynomial Functions of Higher Degree

Before you use the technology, you should understand that:

1. The general n^{th} degree polynomial function can be written as
$$y = a_n x^n + a_{n-1} x^{n-1} + \cdots + a_1 x + a_0$$
The number a_n is called the *leading coefficient* of the polynomial and n is a non-negative integer called the *degree* of the polynomial. If the degree is odd, the graph of the function must cross the x-axis at least once. Equivalently, we say that the polynomial $P(x) = a_n x^n + a_{n-1} x^{n-1} + \cdots + a_1 x + a_0$ has at least one real *zero* when n is odd.

2. We can use the *TI-92* to *approximate* zeros of a polynomial by examining where the graph crosses the x-axis. Then, using zooming techniques, we can refine that approximation. We may also be able to use the *TI-92* to find the zeros *exactly*.

3. The *TI-92* can also help us observe where (in what intervals of x) a function (the y-value) is positive and where it is negative.

Procedures

Procedure #1. In general, as the degree of a polynomial function increases, its graph may exhibit more peaks and valleys (or *turning points*). Also, if the leading coefficient is positive, the graph will rise up to the right; if the leading coefficient is negative, the graph will fall to the right. To see this, we have graphed some polynomial functions in the next eight figures.

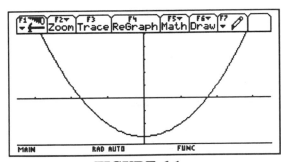

FIGURE 6.1.
Second degree polynomial with $a_n > 0$.

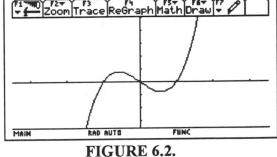

FIGURE 6.2.
Third degree polynomial with $a_n > 0$.

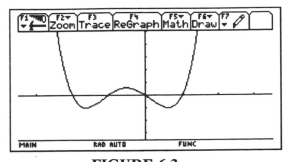

FIGURE 6.3.
Fourth degree polynomial with $a_n > 0$.

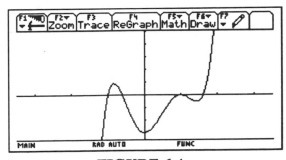

FIGURE 6.4.
Fifth degree polynomial with $a_n > 0$.

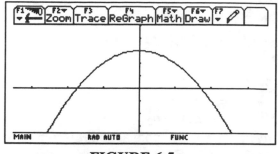

FIGURE 6.5.

Second degree polynomial with $a_n < 0$.

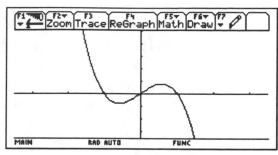

FIGURE 6.6.

Third degree polynomial with $a_n < 0$.

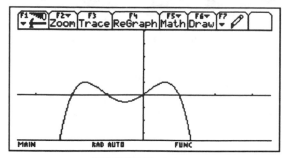

FIGURE 6.7.

Fourth degree polynomial with $a_n < 0$.

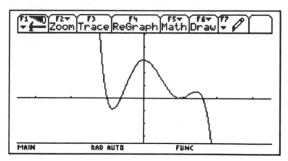

FIGURE 6.8.

Fifth degree polynomial with $a_n < 0$.

Procedure #2. Let's use the *TI-92* to graph $y = x^3 + x^2 - 3x - 3$ and attempt to approximate any real zeros of $x^3 + x^2 - 3x - 3$. (That is, we are looking for values(s) of x that will make the polynomial zero. These are precisely the places where the graph of the polynomial function crosses the x-axis.) The graph appears in Figure 6.9.

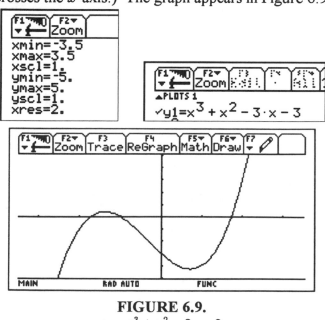

FIGURE 6.9.

$y = x^3 + x^2 - 3x - 3$

It appears that the smallest zero of $x^3 + x^2 - 3x - 3$ falls between -2 and -1. To see how we can refine this approximation, we will use the **TI-92**'s "Zero" command. With the graph displayed, press the [F5] Math key and choose the "2:Zero" option. The **TI-92** will then prompt you for a lower and an upper bound as values that make up an interval that includes the zero. You can cursor to any point or enter values from the keyboard. The values we chose are displayed in Figure 6.10.

Step 1: Choose "2:Zero"

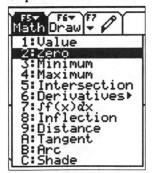

Step 2: Enter Upper and Lower Bounds

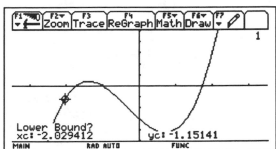

Step 3: The zero is displayed after you press [ENTER]

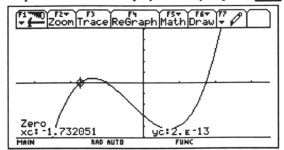

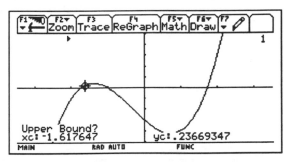

FIGURE 6.10.

Clockwise from upper left: the three steps to find a zero $P(x) = x^3 + x^2 - 3x - 3$ from the GRAPH screen. $P(-1.732051) \approx 0$.

For this example, $P(x) = x^3 + x^2 - 3x - 3$, the **TI-92** can find the *exact* values of all three zeros. We do this using the "1:solve(" command from the [F2] Algebra menu on the HOME screen in Figure 6.11

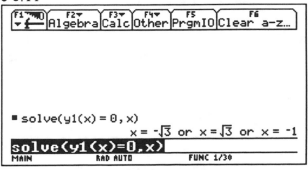

FIGURE 6.11.

If $P(x) = x^3 + x^2 - 3x - 3$, then $P(-\sqrt{3}) = P(\sqrt{3}) = P(-1) = 0$.

Procedure #3. Look carefully at the graph in Figure 6.9. Points on the graph that are above the x-axis are points where the y-coordinate is positive. Similarly, if the curve falls below the x-axis, y values are negative.

Using the **TI-92**'s ability to approximate relative maximum points ("peaks") and relative minimum points ("valleys"), we can use these points to determine the intervals on which the function is increasing (graphically, movement from lower left toward upper right) and where it is decreasing (movement from upper left to lower right). The reader should verify the points highlighted in Figure 6.12.

Approximating the x-value of the peak: Approximating the x-value of the valley:

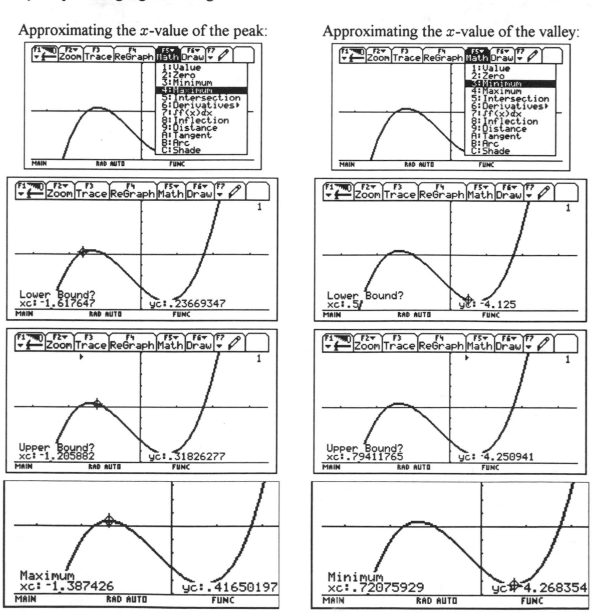

FIGURE 6.12.

It appears that the x-value of the peak is about -1.387426 and the x-value of the valley is about $.72075929$.

We summarize the function's behavior below:

If $x < -\sqrt{3} \approx -1.73$, the function (i.e., the y-values) is negative.   **The function's leftmost x-intercept**
If $-1 < x < \sqrt{3} \approx 1.73$, the function is also negative.   **The function's other x-intercepts**
If $x > \sqrt{3} \approx 1.73$, the function is positive.   **The function's rightmost x-intercept**
If $-\sqrt{3} \approx -1.73 < x < 1$, the function is also positive.   **The function's two leftmost x-intercepts**
If $x < -1.39$, the function is increasing.   **The x-value of the peak**
If $-1.39 < x < 0.72$, the function is decreasing.   **The function's behavior between peak and valley**
If $x > 0.72$, the function is increasing.   **The x-value of the valley**

Some closing comments are in order:

1. In the study of Calculus, we learn ways of finding the values above *exactly*, and need not depend on the approximating techniques of the calculator.

2. In Procedure 2, we used the "solve" command to find the zeros of a third degree polynomial *exactly*. For higher-order polynomials, this cannot always be done. In fact, even for some cubic polynomials, the *TI-92* only approximates answers.[7]

3. Can you find a relationship between the degree of the polynomial and the minimum number of real zeros it must have? Try using your *TI-92* to graph several examples of fourth, fifth, sixth and seventh degree polynomials. Can you find a polynomial with no real zeros? (Try $P(x) = x^6 + 1$ and then find another.) Can you find an odd degree polynomial with no real zeros? See the exercises.

Exercises

In questions 1 through 4, graph the given polynomial function. Also, using the zooming feature, approximate the zeros of the polynomial and the coordinates of any turning points. State where (x- intervals) the function is positive, negative, increasing, and decreasing.
1. $y = 3x^4 + 4x^3$

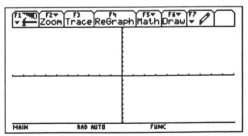

Zeros of the function: For what values of x is y negative?
Turning points: For what values of x is y increasing?
For what values of x is y positive?   For what values of x is y decreasing?

---

[7]For example, try solving $x^3 - 5x^2 + 2x - 1 = 0$. The *TI-92 does* find an approximate answer of $x \approx 4.6134$ but the *exact* answer is $\sqrt[3]{\frac{187 - 9\sqrt{93}}{54}} + \sqrt[3]{\frac{9\sqrt{93} + 187}{54}} + \frac{5}{3}$.

2. $y = 3x^3 - 9x + 1$

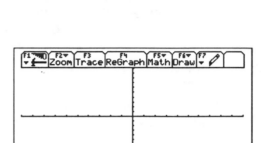

Zeros of the function:

Turning points:

For what values of x is y positive?

For what values of x is y negative?

For what values of x is y increasing?

For what values of x is y decreasing?

3. $y = x^4 - 10x^2 - 11$

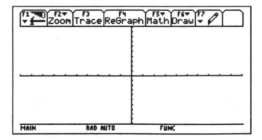

Zeros of the function:

Turning points:

For what values of x is y positive?

For what values of x is y negative?

For what values of x is y increasing?

For what values of x is y decreasing?

4. $y = -x^3 + x^2 + 1$

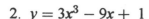

Zeros of the function::

Turning points:

For what values of x is y positive?

For what values of x is y negative?

For what values of x is y increasing?

For what values of x is y decreasing?

5. The graph of $y = x^3 + x^2 - 3x + 3$ is a vertical translation by six units of the graph of $y = x^3 + x^2 - 3x - 3$. (WHY?) Use the *TI-92* in AUTO mode to try to find the one real zero of $y = x^3 + x^2 - 3x + 3$.

6. a) Explain, in your own words, why the zeros of $y = P(x)$ are the same for the function $y = -P(x)$ for any polynomial P.

   b) Are the zeros of $y = P(x)$ the same as the zeros of $y = P(-x)$? Why or why not? Give some examples to substantiate your answer.

   c) Explain why the zeros of $y = P(x)$ must also be zeros of the function $y = |P(x)|$.

7. a) Graph several different examples (make them up yourself) of fifth, sixth, seventh and eighth degree polynomial functions. Include in half of the examples polynomials whose leading coefficient is negative.

   b) True or false: Every odd degree polynomial function must have at least one real zero.

   c) True or false: Every even degree polynomial function must have at least one real zero.

NOTES

Exploration #7:
Inverse Functions

Before you use the technology, you should understand that:

1. Geometrically speaking, two functions are inverses of each other if the graph of one of the functions is the reflection of the other's graph about the line $y = x$.
2. If the point (2, 3) is a point on the graph of $y = f(x)$, then the point (3, 2) must be a point on the graph of its inverse, $y = f^{-1}(x)$.
3. In order for the inverse of $y = f(x)$ to be a *function*, f must be a *one-to-one* function.

Procedures

Procedure #1. Consider the function $f(x) = \frac{3-2x}{4}$. Your mathematics instructor has asked you to find the inverse of this function and you *think* your algebra is correct when you get $f^{-1}(x) = \frac{3-4x}{2}$. To graphically validate that result with the *TI-92* we graph each (linear) function and the line $y = x$ using WINDOW values from a ZoomDec command. If the functions are inverses, their graphs should be "mirror images" about the line $y = x$. Figure 7.1 depicts the three graphs.

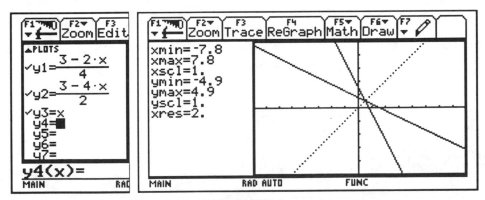

FIGURE 7.1.
The graphs of $y = \frac{3-2x}{4}$, $y = \frac{3-4x}{2}$ and $y = x$ (dotted).

Since the graph of $y = \frac{3-2x}{4}$ and the graph of $y = \frac{3-4x}{2}$ are, in fact, reflections about $y = x$, they are inverse functions. To show this algebraically, we must show that the composition of the original function with its inverse is the identity function, x:

$$f(f^{-1}(x)) = \frac{3 - 2\left(\frac{3-4x}{2}\right)}{4} = x.$$

Procedure #2. If the function is more complicated, the process of solving for the independent variable can be difficult or even impossible. For example, consider $f(x) = \sqrt{2x - 3}$. It can be shown that the inverse is $f^{-1}(x) = \frac{x^2 + 3}{2}$ provided $x \geq 0$. With the *TI-92* program, we can force that restricted domain by using the WITH statement. See Figure 7.2.

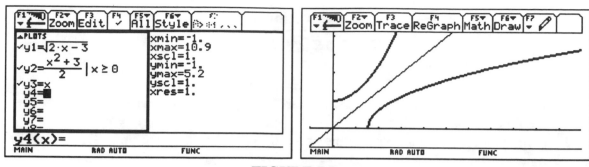

FIGURE 7.2.

$y = \sqrt{2x - 3}$, its inverse and $y = x$.

Again, we see that the graph of $f^{-1}(x) = \frac{x^2+3}{2}$ is the reflection of the graph of f about the line $y = x$.

The graph of a function can be thought of as a collection of points, (p, q). The inverse *relation* of that function is merely the collection of points (q, p). In Figure 7.3, we graph the function $y = f(x) = 4x - \frac{1}{2}x^3$ along with the inverse points $x = f(y) = 4y - \frac{1}{2}y^3$. That latter collection will *not* be a function (Why not?). We use the ***TI-92***-provided DrawInv command to draw the inverse points. To do this, graph $y1(x) = 4x - \frac{1}{2}x^3$ and $y3(x) = x$ first. Then, select "3:DrawInv" from the [F6] Draw menu. Now type $y1(x)$ on the edit line and press [ENTER]. The graph of the inverse will be drawn on the same graph screen.

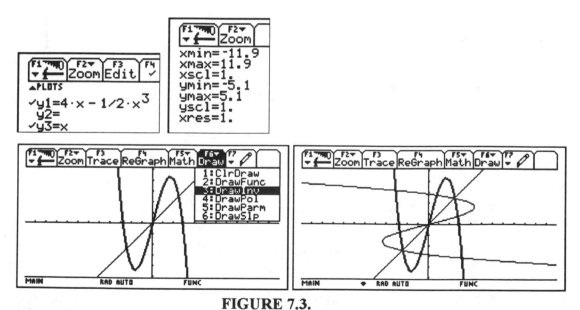

FIGURE 7.3.

The graph of $y = 4x - \frac{1}{2}x^3$ and its inverse relation.

Notice that if we restrict $y = f(x)$ to only those values greater than the peak (where $x \approx 1.63$ in Figure 7.3), then the inverse *will* be a function. Also, in the syntax of the ***TI-92***, we plot the function using the WITH command again. Study Figure 7.4 carefully.

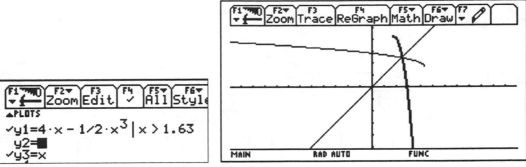

FIGURE 7.4.

$y = 4x - \frac{1}{2}x^3$ ($x > 1.63$) and its inverse.

Exercises

In Exercises 1 through 6, find the inverse for each function. Then, check your work by graphing both the function and the inverse along with the line $y = x$ as in Figure 7.1. It may be helpful to graph $f(x)$ in thick style and $y = x$ in dotted style.

1. $f(x) = 4x$

   $f^{-1}(x) =$

 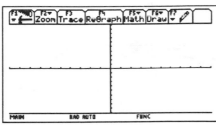

2. $f(x) = \sqrt{4 - x^2}$ $0 \leq x \leq 2$

   $f^{-1}(x) =$

 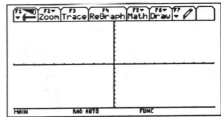

3. $f(x) = \dfrac{4}{x}$

   $f^{-1}(x) =$

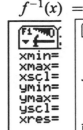 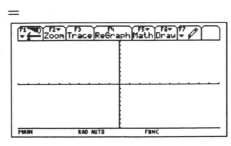

4. $f(x) = \sqrt[3]{x - 1}$

   $f^{-1}(x) =$

 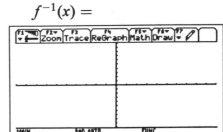

5. $f(x) = x^{3/5}$

6. $f(x) = x^3 + 2$

$f^{-1}(x) =$

$f^{-1}(x) =$

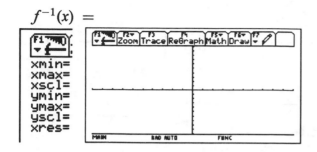

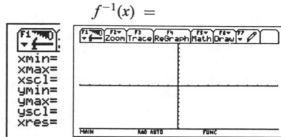

In Exercises 7 and 8 use appropriate scale settings so that you can insure the function (with the appropriate restricted domain) does in fact have an inverse. Be sure to label both the function and its inverse.

7. $f(x) = (x + 3)^2$

8. $f(x) = x^3 - x$

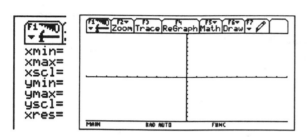

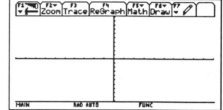

9. Consider $f(x) = \frac{1}{1 + x}$, $x \geq 0$. What is the domain of the inverse of f? Sketch the graph below.

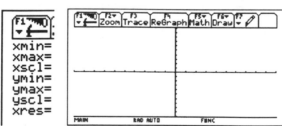

Domain of f^{-1}:   Range of f:

Exploration #8:
Approximating Solutions to Exponential and Logarithmic Equations

Before you use the technology, you should understand that:

1. Although it is a relatively easy task to solve certain contrived exponential or logarithmic equations, such as $2^x = 16$ or $\log(x-2) = 3$, more complicated equations cannot be solved exactly (in closed form) and so must be solved using approximation or graphical techniques. To solve an equation such as $e^{2x} = 1 - x^2$ graphically, we can look to see where the graphs of $y = e^{2x}$ and $y = 1 - x^2$ intersect.

2. Alternatively, we could graph the function $y = e^{2x} - (1 - x^2) = e^{2x} + x^2 - 1$ and determine where its x-intercepts are. Equivalently, that means we are solving $e^{2x} + x^2 - 1 = 0$. In either case, by using a zoom-in technique (graphically) or the *TI-92*'s solve command we can refine the approximation significantly.

Procedures

Procedure #1. We want to find the solutions to the equation $e^{2x} = 1 - x^2$. We observe that $x = 0$ is a solution but after algebraically manipulating the equation without success, we decide to approximate the solution(s). So we graph $y = e^{2x}$ and $y = 1 - x^2$. See Figure 8.1.

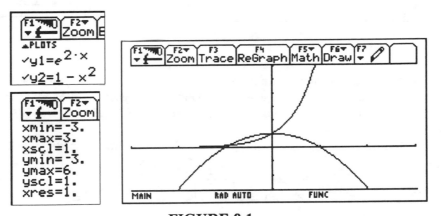

FIGURE 8.1.
The graphs of $y = e^{2x}$ and $y = 1 - x^2$.

It appears that the two functions have two points in common and one of them, $(0, 1)$, can be easily read from the graph and verified algebraically. We are concerned, however, with estimating the coordinates (actually, our task is to find the x-value only) of the other point of intersection. It is appropriate to use the [F3] Trace feature to get a better approximation of the point. Once we have done that, we invoke the *TI-92*'s "5:Intersection" command from the [F5] math menu. See Figure 8.2.

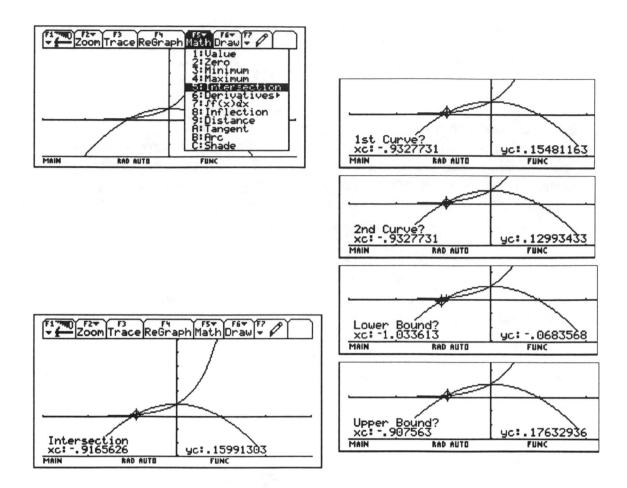

FIGURE 8.2.

After selecting "5:Intersection" from the math menu, the *TI-92* prompts you for four things:
First curve, second curve, lower (left-hand) boundary and upper (right-hand) boundary.
The intersection point is then calculated and displayed as $(-.916526, .15991303)$.

From the HOME screen, we can use the "solve" command to display the approximate solution of $e^{2x} = 1 - x^2$ to more precision. See Figure 8.3.

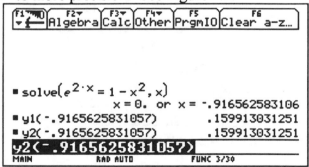

FIGURE 8.3.

The solutions to $e^{2x} = 1 - x^2$ are displayed on the first line of the HOME screen. Also, we check the smaller x-value by finding the corresponding y values for each of $y1(x) = e^{2x}$ and $y2(x) = 1 - x^2$. Indeed, they are equal to twelve decimal places.

We point out here that there are other ways to approximate these solutions. One way would be to author `ZEROS(Y1(X)-Y2(X),X)` from the HOME screen. The two solutions are then listed. For another alternative, consider the *TI-92*'s *very* convenient command called `EXP▷LIST(`. It converts solutions to a list of values. In Figure 8.4 below, we show that alternative.

```
■ exp▶list(solve(y1(x) = y2(x), x), x) → sol
                        {-.916562583106  0.}
■ y1(sol)              {.159913031251  1.}
■ y2(sol)              {.159913031251  1.}
y2(sol)
MAIN        RAD AUTO        FUNC 3/30
```

FIGURE 8.4.

The two solutions to $e^{2x} = 1 - x^2$ are converted to a list variable which we called *sol*. The second and third lines are checks--we have evaluated each x-value in each function.

It should be noted that all five of the following statements are equivalent and the reader should be familiar enough with the vocabulary of mathematics to understand these equivalent statements:

"-0.91656 is an approximate solution to the equation $e^{2x} = 1 - x^2$"

"-0.91656 is an approximate zero of $f(x) = e^{2x} - 1 + x^2$"

"If $f(x) = e^{2x} - 1 + x^2$, then $f(-0.91656) \approx 0$"

"The graphs of $y = e^{2x}$ and $y = 1 - x^2$ intersect at a point whose x-value is about -0.91656."

"The graph of $y = e^{2x} - 1 + x^2$, intersects the x-axis at approximately $(-0.91656, 0)$."

This last interpretation is illustrated in Figure 8.5 below.

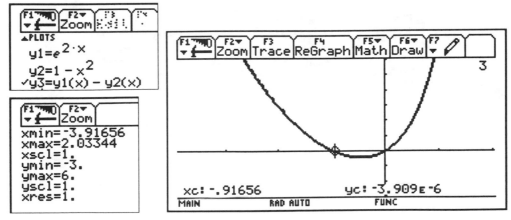

FIGURE 8.5.

The x-intercepts of $y = e^{2x} - 1 + x^2$ are the solutions to $e^{2x} = x^2 - 1$.

Procedure #2. We conclude this exploration by mentioning that the approximation technique discussed here can be applied to *any* functions. Consider the problem of solving the logarithmic equation $\ln(2x - 1) = 3 - \frac{x}{2}$. We will approximate the solution(s) by graphing

$y = \ln(2x - 1)$ and $y = 3 - \frac{x}{2}$ on the same graphics screen and using the **TI-92** intersection command. Figure 8.6 shows that 2.8797309 is the approximate x-value for the equation $\ln(2x - 1) = 3 - \frac{x}{2}$.

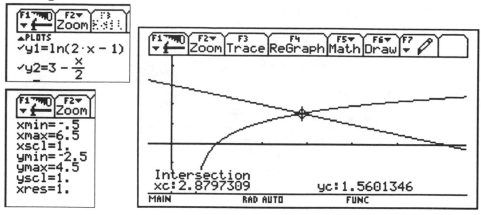

FIGURE 8.6.

The graphs of $y = \ln(2x - 1)$ and $y = 3 - \frac{x}{2}$ after invoking the "5:Intersection" command from the [F5] Math menu and selecting suitable upper and lower bounds.

As before, we relate the graphics screen of Figure 8.6 to the solve command on the HOME screen. In Figure 8.7 the result of using that solve command appears. Again, we check the x values by approximating the corresponding y-values.

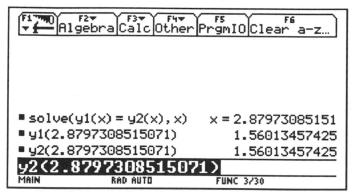

FIGURE 8.7.

The approximate solution to $\ln(2x - 1) = 3 - \frac{x}{2}$ is $x \approx 2.87973085151$.

Exercises

In Exercises 1 through 8

a) Graph each side of the equation as separate functions ($y1(x)$ and $y2(x)$) and use the "5:Intersection" command to approximate the solution.

b) From the HOME screen, use the solve command to substantiate the graphics screen solution in part a).

c) Graph $y1(x) - y2(x)$ and find the x-intercept(s). Verify that they coincide with the solutions found in parts a) and b).

1. $4^{x+3} = 7^x$ Solution(s):

2. $e^x = 3 - 2e^{-x}$ Solution(s):

3. $1 + e^{2x} = \frac{4}{3}$ Solution(s):

4. $\ln(1-x) = \ln 6 - \ln(x+4)$ Solution(s):

5. $\log x = \ln x$ Solution(s):

6. $4 - x^2 = \ln x$ Solution(s):

7. $x^2 = 2^x$ Solution(s):

8. $\frac{e^x + e^{-x}}{e^x - e^{-x}} = 2$ Solution(s):

9. a. Graph these four functions on the same set of coordinate axes:
$$y = \ln x \qquad y = \frac{1}{e}x \qquad y = \frac{1}{e}x + 1 \qquad y = \frac{1}{e}x - 1$$

Using the information in your graph, what can be said about the *number* of solutions to each of the following equations:

b. $\ln x = \frac{1}{e}x$ How many solutions?

c. $\ln x = \frac{1}{e}x + 1$ How many solutions?

d. $\ln x = \frac{1}{e}x - 1$ How many solutions?

10. a. One way to find out how many years it will take for an investment of P dollars at 7.5% annual interest compounded quarterly to triple is to solve the following equation for t:

$$\left(1 + \frac{r}{n}\right)^{nt} = 3 \quad \text{where } r = 0.075 \text{ and } n = 4$$

Solve the equation above for t when $r = 0.075$ and $n = 4$.

b. Solve the problem in part a graphically by plotting $y = \left(1 + \frac{0.085}{4}\right)^{4x}$ and $y = 3$ and finding the abscissa (x-coordinate) of their intersection.

11. When Giselle was presented with the task of approximating the solution to the equation, $e^{0.3-x} = 1 - 2x^3$, she decided to graph the system:

$$y = e^{0.3-x}$$
$$y = 1 - 2x^3$$

    and search for the point of intersection. She obtained the graph below.

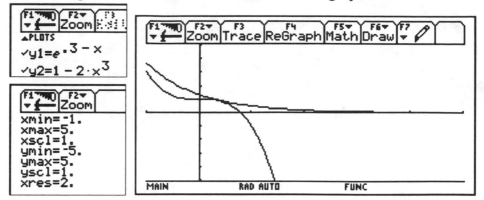

    a. Why was Giselle's guess that the solution was between 0 and 1 a bad decision?

    b. Approximate the solution to the equation $e^{0.3-x} = 1 - 2x^3$ using the graphical approach and better WINDOW settings than Giselle did.

12. a) Graph $y = 1.5^x$ and $y = x^8$ in the same graphing window.
     b) What do you conjecture about the *number* of solutions to the equation $1.5^x = x^8$?
     c) Use the *TI-92* solve command to find the solution(s).

Exploration #9:
The Definition of Derivative

Before you use the technology, you should understand that:

1. The slope of a **secant line** through $y = f(x)$ at $x = a$ and at $x = a + h$, is represented by the *difference quotient*, $\dfrac{f(a+h) - f(a)}{h}$. (The *central* difference quotient is $\dfrac{f(a+h) - f(a-h)}{2h}$ and represents the slope of a secant line through the points on f whose x coordinates are $a + h$ and $a - h$.).

2. The definition of the first derivative of a function $y = f(x)$ at a is given by:

$$f'(a) = \lim_{h \to 0} \frac{f(a+h) - f(a)}{h}$$

Graphically, the first derivative represents the slope of the tangent line at $x = a$ if the limit exists and is finite.

3. The curve has a "smooth" relative maximum or relative minimum where its tangent line is horizontal (and thus where the first derivative equals 0).

Procedures

Procedure #1: Secant lines. To simplify matters somewhat, we will use one function for the entire exploration: $f(x) = x^3 - 6x^2$. We are interested in investigating f around the x value of 2. On the HOME screen, we use the *TI-92*'s ability to allow user-defined functions. In Figure 9.1 we enter f along with formulas for the point-slope form of a straight line and the two-point form of a straight line.[8]

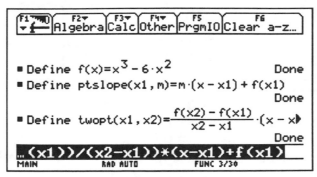

FIGURE 9.1.

$f(x)$ and formulas for the point-slope and two-point form of a straight line are entered on the HOME screen.

---

[8]We use "ptslope" for the name of the function that satisfies the point-slope form of a straight line; it is a function of the point's x-value (x_1) and the line's slope (m). The function "twopt" is the name of the function that we called the two-point form of straight line. It is a function of x_1 and x_2. The y values are all taken care of by $f(x)$.

Now, we enter $f(x)$ as $y1(x)$ in the $Y=$ editor and, since we are interested in the x-value $x_1 = 2$, we examine TABLE[9] values around $x_1 = 2$ in Figure 9.2.

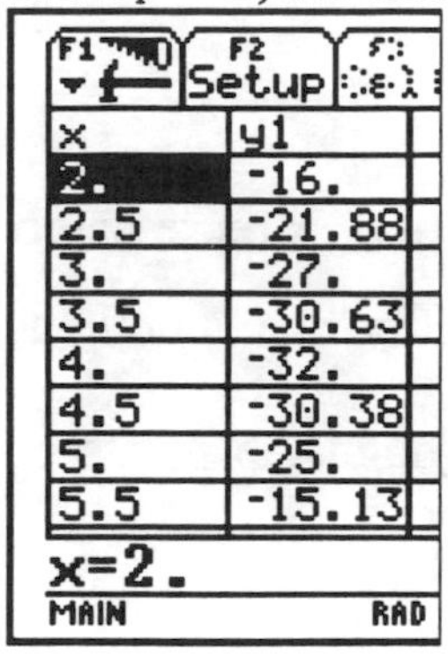

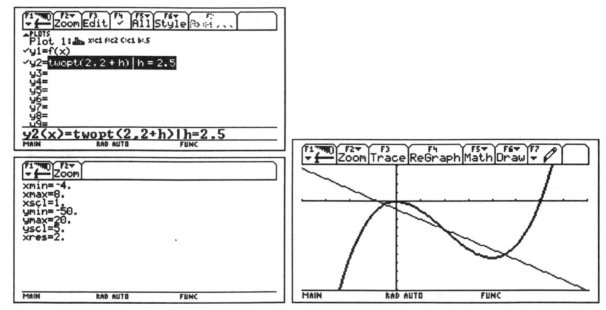

We want to draw a secant line through the two points $(2, f(2))$ and $(2 + h, f(2 + h))$ for some value of h. From the home screen, we find the equation of the line using the two-point form using the *TI-92*'s "with" feature and $h = 2.5$. The equation of the secant line is $y = -5.75x - 4.5$.

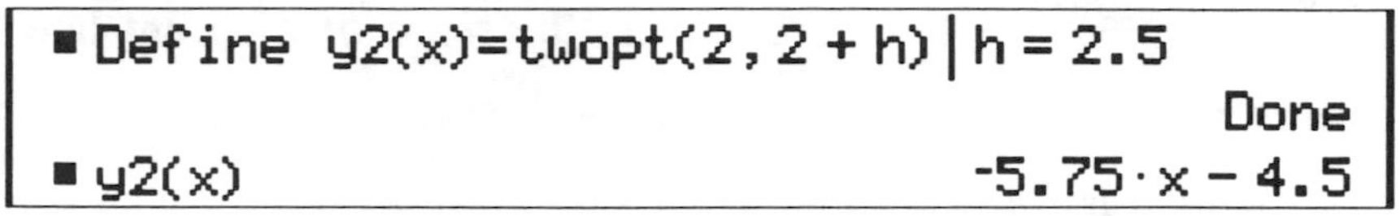

FIGURE 9.2.
Table values around $x_1 = 2$.

The graph of the function and the secant line through the points $(2, -16)$ and $(4.5, -30.375)$ is displayed in Figure 9.3.

FIGURE 9.3.
$y = x^3 - 6x^2$ (drawn in *thick* style) and the secant line through $(2, -16)$ and $(4.5, -30.375)$.

The *TI-92* has a sequence command that allow us to create a list easily. The list we would like to examine is a list of several secant lines, each getting closer and closer to the tangent line at $x_1 = 2$. In Figure 9.4 we issue the command seq(twopt(2,2+h),h,2.5,.5,-.5) to generate a sequence of five secant lines.[10] The list that the *TI-92* returns is $\{ -5.75x - 4.5, -8x, -9.75x + 3.5, -11x + 6, -11.75x + 7.5 \}$. With that high-

---

[9]Remember to use TblSet with tblStart $= 2$, Δtbl $= .5$ and Graph $\leftrightarrow$ Table set to OFF.
[10]If h has been used before, you may have to use DelVar to delete its old definition.

lighted on the HOME screen, we type ◇ Ⓒ to copy it. Then, on the Y = editor screen, we type ◇ Ⓥ to paste it there for $y2$.

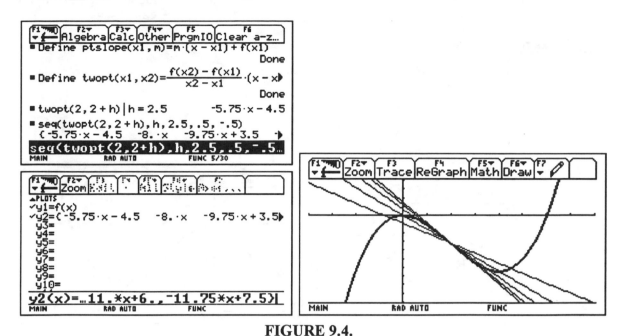

FIGURE 9.4.

Five secant lines drawn closer and closer to $x_1 = 2$.

Procedure #2: Tangent lines. Of course, there are many ways we can find the tangent line drawn to $y = f(x)$ at some point $x = x_1$. In the previous example, we need to allow the point $(2 + h, f(2 + h))$ to approach the point $(2, f(2))$. That is, we need to take the limit as h gets close to zero. This is done in Figure 9.5.

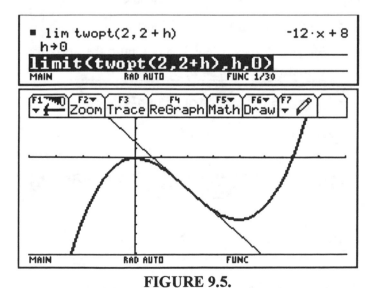

FIGURE 9.5.

The function $y1(x) = x^3 - 6x^2$ and the tangent line drawn to $(2, -16)$, $y2(x) = -12x + 8$.

Once you understand the notion of the limit of a secant line becoming a tangent line, you may want to use the *TI-92*'s built-in functions. From the graph screen, for example, if $y = f(x)$ is graphed you can press the [F5] key for the math menu. Choose "A:Tangent" and when prompted for coordinates, enter a 2 for the x value of the point. See Figure 9.6.

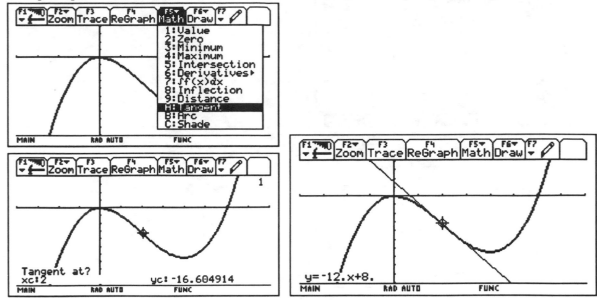

FIGURE 9.6.

Finding the equation of a tangent line from the graph screen.

The derivative, defined as the limit of the difference quotient as h tends to zero, really represents a certain **rate of change**. As h gets close to zero, the derivative is thought of as the instantaneous rate of change of the function with respect to x. Finally, we mention that the *TI-92* will find the symbolic form of the first derivative -- that is, it can express the value of the slope of the tangent line at *any* point as a function of x.

The HOME screen displayed in Figure 9.7 calculates $f'(x) = \dfrac{dy}{dx} = 3x^2 - 12x$. We create a function called *tline* (for "tangent line") to compute the equation of the tangent line at any point x_1. Then, we evaluate *tline* at $x = 2$.

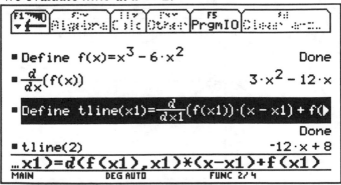

FIGURE 9.7.

The function to evaluate the equation of a tangent line, *tline*, is:
tline(x1) = d(f(x1),x1)*(x − x1)+f(x1).

The keystrokes for entering the derivative directly involve pressing [2nd] then the numeric [8] key.

Exercises

In questions 1 through 7, a) find the equation of the secant line that goes through the points $(x_1, f(x_1))$ *and* $(x_2, f(x_2))$; *b) find the equation of the line drawn tangent to* $y = f(x)$ *that passes through* $(x_1, f(x_1))$; *c) construct a graph of the function along with the tangent line.*

1. $f(x) = x^3 - 3x$ $\qquad\qquad$ $x_1 = -2$ $\qquad\qquad$ $x_2 = -1$

2. $f(x) = x^{1/3}$ $\qquad\qquad$ $x_1 = -2$ $\qquad\qquad$ $x_2 = 2$

3. $f(x) = x^{2/3}$ $\qquad\qquad$ $x_1 = -1$ $\qquad\qquad$ $x_2 = 1$

4. $f(x) = \dfrac{1}{9 - x^2}$ $\qquad\qquad$ $x_1 = -2$ $\qquad\qquad$ $x_2 = 2$

5. $f(x) = \sin x$ $\qquad\qquad$ $x_1 = \frac{\pi}{4}$ $\qquad\qquad$ $x_2 = \pi$

6. $f(x) = e^{-x^2}$ $\qquad\qquad$ $x_1 = -1$ $\qquad\qquad$ $x_2 = 1$

7. $f(x) = e^{-x^2}$ $\qquad\qquad$ $x_1 = 0$ $\qquad\qquad$ $x_2 = 1$

8.   Repeat the work of procedure #1 but this time use the **central difference quotient,** $\dfrac{f(a + h) - f(a - h)}{2h}$. Use $f(x) = x^3 - 6x^2$, $a = 2$, and various values of h.

9.   A straight line is said to be a **normal** line to a curve at a point $(x_1, f(x_1))$ if it is perpendicular to the tangent line at that point. That means that the slope of the normal line must be $-\dfrac{1}{f'(x_1)}$. (Why?)   Consider the function $f(x) = \ln x$ along with its tangent and normal lines drawn to $x = 2$ below:

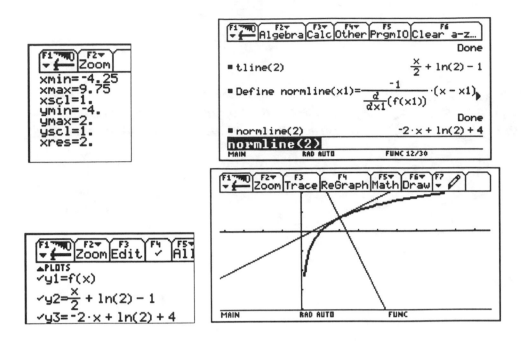

Find the equation of the normal line drawn to $f(x) = x^3 - 6x^2$ at $x = 2$.

10.   The derivative represents the slope of a tangent line drawn to a curve at a particular
point. Therefore, the derivative is certainly related to the shape of a curve. The graph
below is the result of graphing 13 tangent lines drawn to the x values $-3, -2.5, -2,$
$-1.5, ...2.5, 3$. They were generated on the *TI-92* by graphing the sequence:

$$\text{seq}(-2k(x-k)+4-k^2, k, -3, 3, .5)$$

Notice that the shape of the curve $y = 4 - x^2$ appears even though only straight lines
were graphed! (It is not hard to understand why the family of these straight lines is
called the **envelope** of $y = 4 - x^2$.)

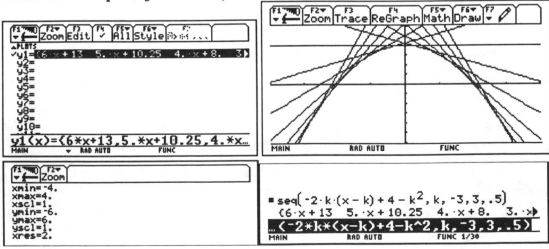

Explain what $-2k(x-k)+4-k^2$ represents and repeat the process for 21 straight
lines (let k go from -4 to 4 in increments of 0.2 units). (This may take a few
minutes.)

Exploration #10:
Relative and Absolute Extrema

Before you use the technology, you should understand that:

1. If f is a continuous function and $f'(x)$ changes from negative to positive at c, then $f(c)$ is a relative minimum; if $f'(x)$ changes from positive to negative at c, then $f(c)$ is a relative maximum. We refer to relative maxima and relative minima collectively as *relative extrema*. Some texts use the word *local* instead of *relative*.

2. It is important to note that if a function f is differentiable at $x = c$ and has a relative maximum (or minimum) at $x = c$, then the graph of $y = f'(x)$ must cross the x-axis at c (because $f'(c) = 0$). [Of course, if f is *not* differentiable, there may still be a relative extremum.]

3. A graph will be concave upward when the second derivative is positive and will be concave downward when the second derivative is negative. Where a continuous function, f, changes from concave upward to concave downward is called a *point of inflection* of f.

Procedures

Procedure #1: Graphing a function and its derivative. Consider the function graphed in Figure 10.1, $f(x) = -3x^4 + 4x^3 + 3x^2 - 6x$. It appears that the function has an absolute maximum value somewhere between $x = -1$ and $x = 0$. The question is what is that exact x value and are there any other extrema (either absolute or relative) for this function.

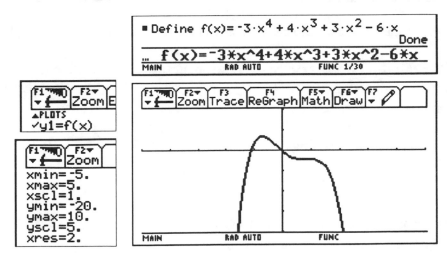

FIGURE 10.1.

There is an obvious (absolute) extreme value between $x = -1$ and $x = 0$. What about between 0 and 2?

Later we will zoom-in to get a better idea of the behavior of this function between 0 and 2. For now, we choose to use the same graphing window to also plot $y = f'(x)$. This is done in Figure 10.2.

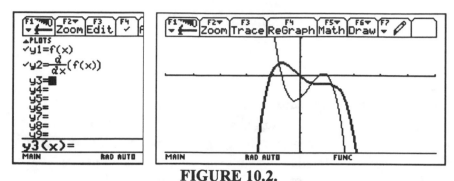

FIGURE 10.2.

Aha! Since the first derivative crosses the x axis twice between 0 and 2,
there may be other extrema in this interval.[11]

Notice also from the graph in Figure 10.2, that the graph of $y = f(x)$ is increasing where the graph of its first derivative is positive and it is decreasing where the first derivative is negative. A ZoomBox in Figure 10.3 emphasizes the behavior of the function between 0 and 2. (We've turned $y2(x)$ off.)

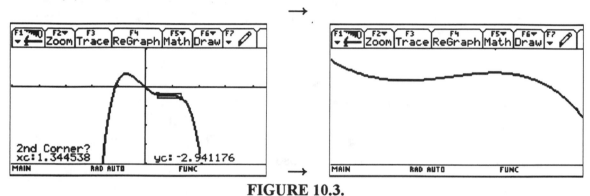

FIGURE 10.3.

There appears to be a relative minimum *and* a relative maximum in the interval $[0, 2]$.

Procedure #2: Finding *exact* extrema. By defining the first derivative (we use $f1$) and the second derivative (defined as the function $f2$ in Figure 10.4) on the HOME screen, we can find the exact x-values that set the first derivative equal to zero. Then, by evaluating $f2$ at each of the three values for which $f'(x) = 0$, we can apply the second derivative test to try to determine whether they are maxima or minima (or neither). See Figures 10.4 and 10.5.[12]

---

[11]A helpful suggestion: Use the "thick" style for the function and another style (say, "line") for the derivative. Recall that this is done from the Y = editor by pressing the [F6] key.

[12]There are other ways to find maximum and minimum values. From the graph screen choose [F5] math.

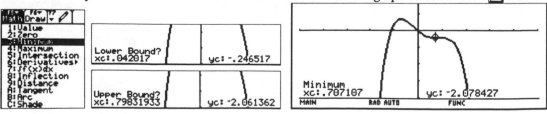

```
F1▼  F2▼   F3▼   F4▼   F5     F6
▼ ┌─ Algebra Calc Other PrgmIO Clear a-z...

■Define f(x)=-3·x⁴+4·x³+3·x²-6·x
                                    Done
■Define f1(x)=d/dx(f(x))            Done

■Define f2(x)=d²/dx²(f(x))          Done

Define f2(x)=d(f(x),x,2)
MAIN      RAD AUTO      FUNC 3/30
```

```
F1▼  F2▼   F3▼   F4▼   F5     F6
▼ ┌─ Algebra Calc Other PrgmIO Clear a-z...

■Define f1(x)=d/dx(f(x))           Done

■Define f2(x)=d²/dx²(f(x))         Done

■solve(f1(x) = 0, x)
        x = √2/2 or x = -√2/2 or x = 1

solve(f1(x)=0,x)
MAIN      RAD AUTO      FUNC 4/30
```

FIGURE 10.4.

On the left, the three definitions for the function and its first and second derivatives. On the right, we find that the three x-values to examine for extrema are $-\frac{\sqrt{2}}{2}$, $\frac{\sqrt{2}}{2}$, and 1 by setting the first derivative equal to 0.

```
■solve(f1(x) = 0, x)
        x = √2/2 or x = -√2/2 or x = 1

■{ f2(-√2/2)  f2(√2/2)  f2(1) }
              {-28.9706  4.97056  -6.}
```

```
■{ f(-√2/2)  f(√2/2)  f(1) }
  { -2³ᐟ²/2 + 3·√2 + 3/4   2³ᐟ²/2 - 3·√2 + 3/4  ▶

■{ f(-√2/2)  f(√2/2)  f(1) }
              {3.57843  -2.07843  -2.}
```

FIGURE 10.5.

On the left, we approximate each of the three values to see if $\frac{d^2y}{dx^2}$ is positive (minimum) or negative (maximum). On the right, we evaluate the function at each of the 3 values to find the y values of the points.[13]

The *TI-92* command EXP▷LIST(ANS(1),X) is a useful command to convert the solutions to list form for easy evaluation. In Figure 10.6 we store that list in the variable called *sol* and then evaluate *sol* for $f2$ to quickly show us the sign of the second derivative for each x-value. Also, $f(sol)$ evaluates the original function at each x value.

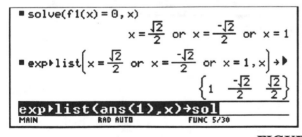

```
■solve(f1(x) = 0, x)
        x = √2/2 or x = -√2/2 or x = 1

■exp▶list( x = √2/2 or x = -√2/2 or x = 1,x )▶
                        { 1  -√2/2  √2/2 }

exp▶list(ans(1),x)→sol
MAIN      RAD AUTO      FUNC 5/30
```

```
■f2(sol)
   {-6.   -28.9705627485   4.97056274848}
■f(sol)
   {-2.   3.57842712475   -2.07842712475}

f(sol)
MAIN      RAD AUTO      FUNC 7/30
```

FIGURE 10.6.

Left: we use the EXP▷LIST(ANS(1),X) command and store the result in variable *sol*.
Right: notice how the second derivative and the function are evaluated.

Now, we can also make a statement about where (on which intervals) the curve is increasing and where it is decreasing:

---

[13]Caution: The 3 points are *exactly* $\left(-\frac{\sqrt{2}}{2}, \frac{8\sqrt{2}+3}{4}\right)$, $\left(\frac{\sqrt{2}}{2}, \frac{-8\sqrt{2}+3}{4}\right)$, and $(1, -2)$. The *TI-92* expresses $\frac{8\sqrt{2}+3}{4}$ rather awkwardly as $\frac{-2^{3/2}}{2} + 3\sqrt{2} + 3/4$.

Increasing: $-\infty < x < -\frac{\sqrt{2}}{2}$ and $\frac{\sqrt{2}}{2} < x < 1$

Decreasing: $-\frac{\sqrt{2}}{2} < x < \frac{\sqrt{2}}{2}$ and $1 < x < \infty$

Procedure #3: Absolute vs. Relative Extrema. There are two cases where we may be asked to find the absolute extrema of a continuous function: either on its entire domain or on some closed interval. The candidates in the former case are the relative extrema and $\pm\infty$; in the closed interval case, we examine relative extrema and the endpoints of the interval. The table below examines our function $f(x) = -3x^4 + 4x^3 + 3x^2 - 6x$ on a variety of intervals.

Interval:	x-value of Abs. Max	Value of Abs. Max	x-value of Absolute Min	Value of Absolute Min
$(-\infty, \infty)$	$-\frac{\sqrt{2}}{2} \approx -0.707$	$\frac{8\sqrt{2}+3}{4} \approx 3.58$	None[14]	$-\infty$
$[\frac{\sqrt{2}}{2}, 1.1]$	1	-2	$\frac{\sqrt{2}}{2} \approx 0.707$	$\frac{-8\sqrt{2}+3}{4} \approx -2.08$
$[\frac{\sqrt{2}}{2}, 1, 5]$	1	-2	1.5	-3.9375
$[-5, 5]$	$-\frac{\sqrt{2}}{2} \approx -0.707$	$\frac{8\sqrt{2}+3}{4} \approx 3.58$	-5	-2270

TABLE 10.1.

Examining absolute extrema on intervals for $f(x) = -3x^4 + 4x^3 + 3x^2 - 6x$.

Finally, we conclude with an example of a function that is *not* continuous on some interval. We want to examine a function that has an asymptotic discontinuity and we choose $g(x) = \frac{x^2}{x^2-9}$ From the graph in Figure 10.7, we can see a relative maximum at $x = 0$.

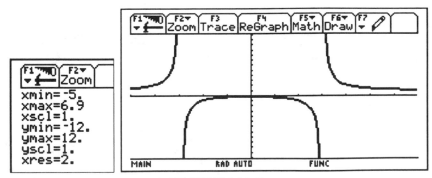

FIGURE 10.7.

Finding absolute extrema over intervals containing ± 3 means checking asymptotes. See Table 3.2.

---

[14]The *TI-92* will actually return the values $\pm\infty$ if you use the minimum and maximum functions from the calculus menu:

∎ fMin(f(x),x)	x = ∞ or x = -∞
∎ fMax(f(x),x)	$x = \frac{-\sqrt{2}}{2}$

Interval:	x-value of Absolute Max	Value of Absolute Max	x-value of Absolute Min	Value of Absolute Min
$[-2, 1]$	0	0	-2	-0.8
$[-1, 2]$	0	0	2	-0.8
$[-4, 2]$	None	∞	None	$-\infty$
$[4, 6]$	4	16/7	6	4/3

TABLE 10.2.

Examining absolute extrema on intervals for $g(x) = x^2/(x^2 - 9)$.

Procedure #4: Concavity and points of inflection. To examine concavity, we return to our original example, $f(x) = -3x^4 + 4x^3 + 3x^2 - 6x$. When $f''(x) > 0$, the curve is concave upward; when $f''(x) < 0$, it is concave downward. We examine the graphs of $y = f(x)$ and $y = f''(x)$ in Figure 10.8. Keep in mind that places where $f''(x) > 0$ (that is above the x-axis) are places where the original function $f(x)$ is concave up. To find the values of these points of inflection, we employ the *TI-92*'s "solve(" command.

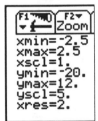

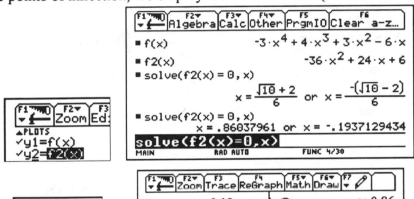

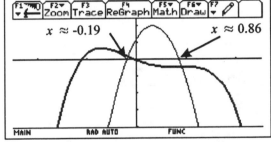

FIGURE 10.8.

Two points of inflection: $(\frac{\sqrt{10}+2}{6}, f(\frac{\sqrt{10}+2}{6})) \approx (.86, -2.04)$ and $(\frac{-\sqrt{10}+2}{6}, f(\frac{-\sqrt{10}+2}{6})) \approx (-0.19, 1.24)$. The approximate values are obtained by pressing $\boxed{2^{nd}}$ $\boxed{\text{ENTER}}$.

Exercises

In exercises 1 through 5, find the absolute extrema and relative extrema for the intervals indicated. Find the intervals where the curve is increasing and where it is decreasing. Also, find the intervals where the curve is concave upward, the intervals where it is concave downward, and points of inflection.

1. $f(x) = 9x - x^3$ on the interval $[-4, 4]$.

   Relative extrema: _____ Absolute Extrema: _____

   Interval(s) where the curve is increasing: _____

   Interval(s) where the curve is decreasing: _____

   Interval(s) concave down: _____   Interval(s) concave up: _____

   Point(s) of inflection: _____

2. $f(x) = \dfrac{x}{x^2 + 1}$ on the interval $[-4, 4]$.

   Relative extrema: _____ Absolute Extrema: _____

   Interval(s) where the curve is increasing: _____ decreasing: _____

   Interval(s) concave down: _____   Interval(s) concave up: _____

   Point(s) of inflection: _____

3. $f(x) = \dfrac{x}{x^2 + 1}$ on the interval $(-\infty, \infty)$.

   Relative extrema: _____ Absolute Extrema: _____

   Interval(s) where the curve is increasing: _____ decreasing: _____

   Interval(s) concave down: _____   Interval(s) concave up: _____

   Point(s) of inflection: _____

4. $f(x) = \dfrac{x^2 + 1}{x}$ on the interval $[-4, 4]$.

    Relative extrema: _____ Absolute Extrema: _____

    Interval(s) where the curve is increasing: _____ decreasing: _____

    Interval(s) concave down: _____ Interval(s) concave up: _____

    Point(s) of inflection: _____

5. $f(x) = x^{2/3}$ on the interval $[-4, 4]$.

    Relative extrema: _____ Absolute Extrema: _____

    Interval(s) where the curve is increasing: _____ decreasing: _____

    Interval(s) concave down: _____ Interval(s) concave up: _____

    Point(s) of inflection: _____

NOTES

Exploration #11:
Newton's Method

Before you use the technology, you should understand that:

1. The zero of a function, say $f(x) = x^4 + x^3 - 2x^2 - 3x - 3$, is a value of x where the graph of $y = f(x)$ touches or crosses the x-axis, that is, where $f(x) = 0$. Newton's method gives us a way of approximating those zeros by:

   a) using an initial value, called a *guess*, to find a next value

   b) calculating that next value according to

$$x_{n+1} = x_n - \frac{f(x_n)}{f'(x_n)}$$

   c) repeating this process (that is, *iterations* are performed) until an approximation is "close enough", that is, within some predetermined tolerance.

2. The method can be used to approximate irrational numbers by realizing that, for example, $\sqrt{5}$ is a zero of the function $f(x) = x^2 - 5$ or π is a zero of the function $g(x) = \sin x$.

3. The above process fails if the sequence of x_i's does not approach a zero of f or if an x value occurs for which the derivative, f', is zero.

Procedures

Procedure #1. To begin, we consider the function $f(x) = x^4 + x^3 - 2x^2 - 3x - 3$. We will use $g(x)$ to represent the first derivative of f (that is, $g(x) = f'(x)$). Also, $newt(x)$ is the iteration function: $newt(x_n) = x_n - \dfrac{f(x_n)}{g(x_n)}$. We define f, g, and $newt$ in the HOME screen as shown in Figure 11.1.

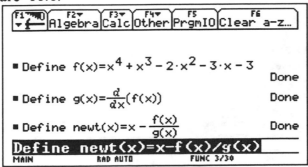

FIGURE 11.1.
Preparation for Newton's Method: three functions are entered using the "Define" command.

Since we are interested in finding the zeros of this function and we need an initial "guess", it is appropriate that we graph $y = f(x)$ before proceeding. Of course, the zeros of f coincide precisely with the x-intercepts of the graph of $y = f(x)$. Figure 11.2 displays the graph of f. Observe that the two real zeros of f occur between -2 and -1 and between 1 and 2.

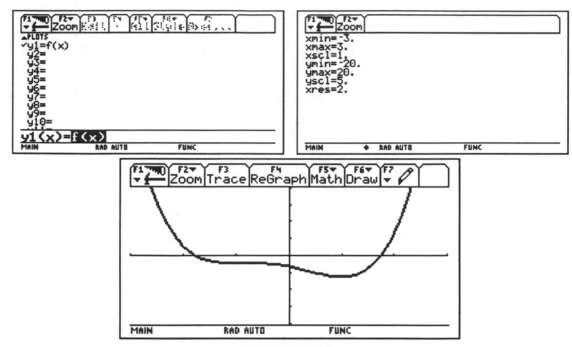

FIGURE 11.2.

$f(x)$ appears to have two real zeros. One of them is "near" $x = -2$ and the other is near $x = 2$.

We will focus on the positive real solution between $x = 1$ and $x = 2$. For this reason, we choose an initial *guess* of $x_1 = 2$. Finding $newt(2)$ will produce the first in a series of x values that we hope converge to the actual zero. For these calculations, we put the **TI-92** into "Approximate" mode with "Display Digits" set at FLOAT 8. The value of $newt(2) \approx$ 1.7878788. In Figure 11.3 the results of the first five iterations are displayed.

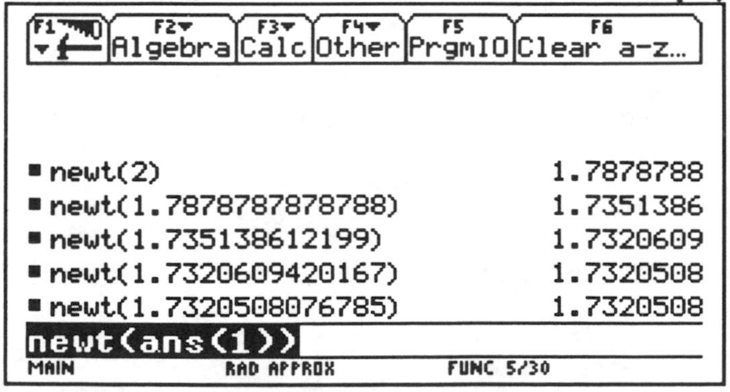

FIGURE 11.3.[15]

The value of x_6, the result of the fifth iteration of Newton's Method, is about 1.7320508.

---

[15]The first line is entered as newt(2). For the second line, type newt(ans(1)) and then press ENTER four times to get the bottom four lines of iterations.

Procedure #2. On the *TI-92*, there is a natural way to look at a function like $newt(x)$. To visualize the sequence of the iterations of Newton's Method, we change the graph MODE to "SEQUENCE".

Now, we define the sequence $u1(n)$ to be $newt(n-1)$ and the initial value of $u1$, that is, $ui1$, is assigned the value 2. Then the sequence can be listed on the home screen (see Figure 11.4) or it can be viewed on the graphing screen (see Figure 11.5).

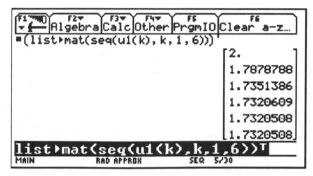

FIGURE 11.4.

With $u1$ defined in the Y = screen (upper left), we can create a list ("seq") of six x values (lower left).[16]

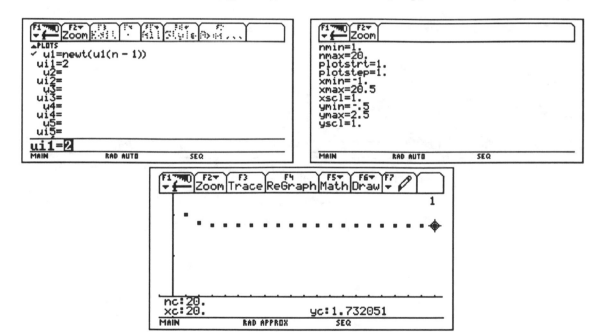

FIGURE 11.5.

A sequence graph of "$newt(x)$" using Trace.[17]

---

[16]On the right, we use a convenient (but tricky) device to display our sequence vertically. The list ▷ mat() procedure (available from the MATH LIST submenu) converts lists into matrices by filling rows with values from the list (sequence). After creating our 1 row matrix, we take its transpose with the ^{T} operator (available from the MATH MATRIX submenu) to produce a column vector of values from our sequence.

[17]You may want to read the section on sequence plotting in the *Overview* portion of this text on pages 37-41.

Another natural way to observe the values of *newt* is through the TABLE screen. The table values are truncated in the table, but when you highlight an entry, it is displayed to 14 digits of precision on the edit line. See Figure 11.6 for our table created by using $\boxed{\diamond}$ $\boxed{\text{TblSet}}$ with tblStart $= 1.0$ and Δtbl $= 1.0$.

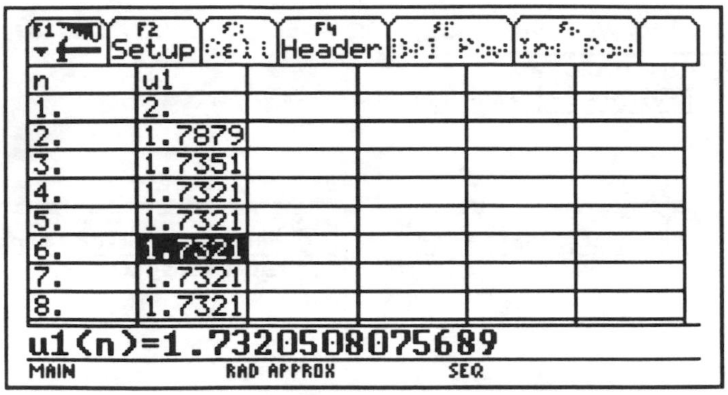

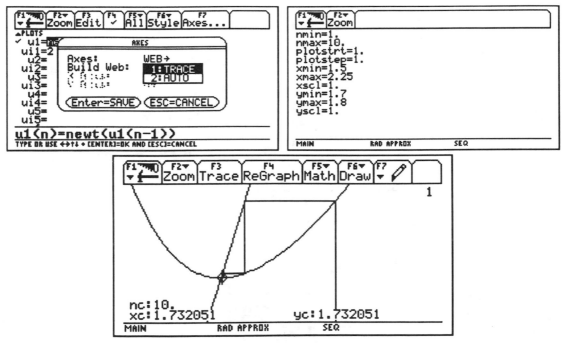

FIGURE 11.6.
A look at the TABLE of iterated values of Newton's Method.

Finally, we use the *TI-92* WEB option (chosen under $\boxed{\text{F7}}$ Axes on the Y $=$ editor screen) to see yet another look at the convergence of the iteration.

FIGURE 11.7.
A WEB graph. The straight line is $y = x$; the curved line is $y = newt(x)$ and its iterations occur as we repeatedly press the right cursor key in trace mode. This graph was "traced" 10 times ("nc:10"). We observe convergence to the value ≈ 1.732051.

Exercises

1. For the function in Procedure #1, $f(x) = x^4 + x^3 - 2x^2 - 3x - 3$, we used Newton's Method to approximate the zero nearest to $x = 2$.
   a. Use an initial guess of $x_1 = -2$ to approximate the other zero.
   b. Which zero does an initial guess of $x_1 = 0$ begin approaching?

2. It happens that the zeros of the function that we introduced in Procedure #1, $f(x) = x^4 + x^3 - 2x^2 - 3x - 3$, can be found *exactly*. Find them using the *TI-92*'s "solve(" command and compare to the values we approximated in this exploration. [Be sure to put the *TI-92* back into AUTO mode (and out of APPROX mode).]

In questions 3 through 7, for each function and initial guess, estimate a zero using three and then five iterations of Newton's Method. Be sure to be in approximate mode and express your answers using eight digits of precision.

Function	Initial Guess	Value of root after Three Iterations	Value of root after Five Iterations
3. $f(x) = x^3 + x^2 - 7x + 5$	$x_1 = 2$		
4. $f(x) = x^3 + x^2 - 7x + 5$	$x_1 = -4$		
5. $f(x) = x^3 - 15x^2 + 63x - 64$	$x_1 = 8$		
6. $f(x) = x^3 - 15x^2 + 63x - 64$	$x_1 = 1$		
7. $f(x) = \sqrt{2x+1} - \sqrt{x+4}$	$x_1 = 5$		

8. Approximate $\sqrt{70}$ by using five iterations of Newton's method to approximate a zero of $f(x) = x^2 - 70$; use an appropriate initial guess.

9. Approximate e by using five iterations of Newton's Method to approximate a zero of $f(x) = -1 + \ln x$; use $x_1 = 3$ as an initial guess.

10. Approximate $\sqrt[3]{100}$ by using five iterations of Newton's Method, an appropriate function, and an initial guess of 4.

11. a) Consider the function $f(x) = 3x^4 - 8x^3 + 6x^2 - 24x + 1$. What happens when Newton's method is applied in the first iteration with an initial guess of 2? Why?

    b) Can you approximate a zero of f with a different initial guess? Try 1.

12. In this exercise, we attempt to explain some of the theory behind Newton's Method. The iteration process can be explained as follows: Draw a tangent line at the point on the function at the initial value, namely $(x_1, f(x_1))$. Where this line crosses the x axis, $x = x_2$, is often closer to the zero of the function than x_1 is. Then the tangent line is drawn to the curve at the point $(x_2, f(x_2))$, and so on.

We graph a family of three of these tangent lines below. (This may take a while!)

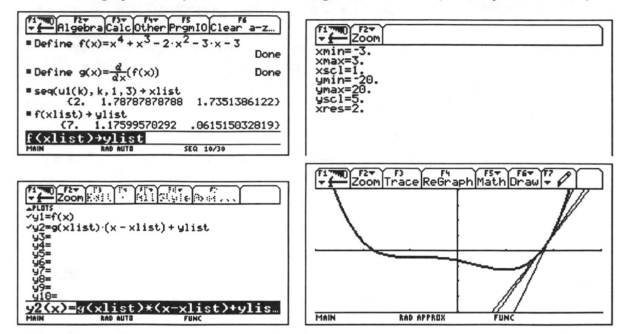

Next, we do a ZoomBox to concentrate more on the area around the x-intercepts.

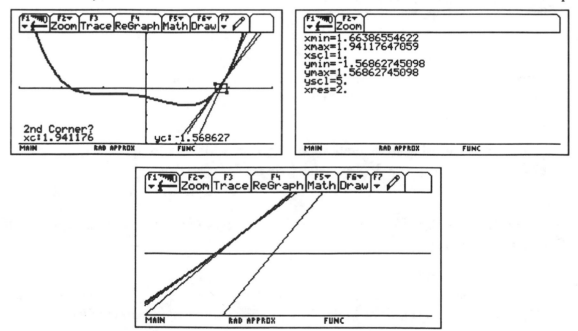

Repeat this process for the zero of the function near $x = -2$.

13. Consider $f(x) = (x - 0.5)^{1/3}$ with an initial guess of 1. What occurs in the application of Newton's Method? [HINT: Find $newt(1)$, $newt(newt(1))$, $newt(newt(newt((1)))$, etc.]

14.   Explain the graph below.

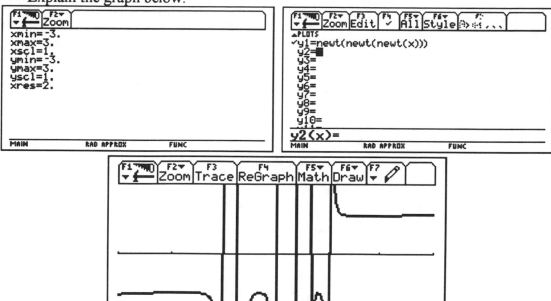

HINT: Notice that the graph has two flattened plateaus. What are the y-values of these plateaus? Notice that these values compare with f's two zeros. The vertical lines occur at places that represent bad initial guesses.

NOTES

Exploration #12:
Simpson's Rule

Before you use the technology, you should understand that:

1. For this exploration, we will assume $y = f(x)$ is a positive, continuous function in some closed interval $a \leq x \leq b$ and that n is a positive even integer. Then the area between $y = f(x)$ and the x-axis between $x = a$ and $x = b$ is given by

$$\int_a^b f(x)\, dx \approx \frac{h}{3}[f(x_0) + 4f(x_1) + 2f(x_2) + 4f(x_3) + \cdots + 2f(x_{n-2}) + 4f(x_{n-1}) + f(x_n)]$$

where $h = \dfrac{b-a}{n}$, the width of each subinterval. This approximation to $\int_a^b f(x)\, dx$ is known as Simpson's Rule (or the Parabolic Rule).

Simpson's Rule is based on the fact that any three non-collinear points determine a unique parabola which passes through them. We can partition $[a, b]$, using groups of three adjacent points on $f(x)$ and approximate $\int_a^b f(x)\, dx$ by the area under the parabolic segments that approximate $f(x)$. Note that since we need three adjacent points, our partition must have an even number of subintervals.

2. The error, S_{err}, in this approximation is given by $0 \leq S_{err} \leq \frac{(b-a)^5 M}{180 n^4}$ where M is the maximum value for $\left| f^{(4)}(x) \right|$ on the interval $a \leq x \leq b$.

Procedures

Procedure #1. Our first goal is to *visualize* Simpson's Rule. We begin with a function that has a lot of curvature, $f(x) = -x^6 + 20x^2 + 10x + 40$ over the interval $-2 \leq x \leq 2$. This function, using a suitable scale, is graphed in Figure 12.1

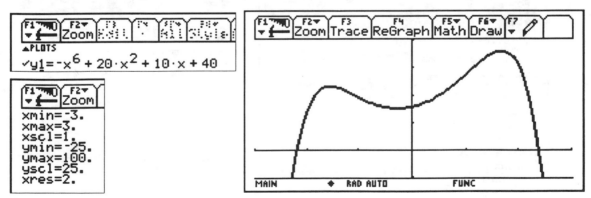

FIGURE 12.1.
$$f(x) = -x^6 + 20x^2 + 10x + 40$$

Let's use $n = 4$ to begin. Now, we need to divide the interval $-2 \leq x \leq 2$ into four sub-intervals, $-2 = x_1 \leq x_2 \leq x_3 \leq x_4 \leq x_5 = 2$. In order to find the parabola (whose equation is $y = ax^2 + bx + c$) that passes through the points (x_1, y_1), (x_2, y_2), (x_3, y_3), we must find the values of a, b and c in the system of equations

$$y_1 = ax_1^2 + bx_1 + c$$
$$y_2 = ax_2^2 + bx_2 + c$$
$$y_3 = ax_3^2 + bx_3 + c$$

We leave it to the reader to verify that, using matrices, the values of a, b, and c are given by

$$\begin{bmatrix} a \\ b \\ c \end{bmatrix} = \begin{bmatrix} x_1^2 & x_1 & 1 \\ x_2^2 & x_2 & 1 \\ x_3^2 & x_3 & 1 \end{bmatrix}^{-1} \begin{bmatrix} y_1 \\ y_2 \\ y_3 \end{bmatrix} \tag{1}$$

Now that parabola would have to be graphed over the interval $x_1 \le x \le x_2$ and then the process repeated for each of the parabolas (in this case there are 2 parabolas). Thankfully, the *TI-92* can do much of this for us. In the listing in Figure 12.2, notice that the following line of code: "xp^(-1)*yp $\rightarrow$ parab" corresponds to matrix equation (1), above. Also, the actual parabolas are graphed as a defined function, $g(x)$, with the line "DrawFunc g(x)".

```
F1    F2     F3   F4  F5      F6
  Control I/O Var Find... Mode
:simpson1()
:Prgm
:@ NOTE  Function must be y1
:ClrIO  @Clears the output screen
:ClrDraw  @Clears previous objects
:randMat(3,3)→xp
:randMat(3,1)→yp
:randMat(3,1)→parab
:Input "Enter left endpoint, a",a
:Input "Enter right endpoint, b",b
:Input "Enter number of EVEN divisions,
 n",n

:(b-a)/n→h   @ h is the width of subdiv.
:seq(a+k*h,k,0,n)→xlist      @ x-values
:y1(xlist)→ylist             @ y-values
:For k,1,n-1,2
:@ "r" loop is done for each parabola
:For r,1,3
:xlist[r+k-1]^2→xp[r,1]
:xlist[r+k-1]→xp[r,2]
:1→xp[r,3]
:ylist[r+k-1]→yp[r,1]
:EndFor
:xp^(-1)*yp→parab      @ Calculate coeffs

:Define g(x)=Func
:If x>a+(k-1)*h and x<a+(k+1)*h Then
:Return parab[1,1]*x^2+parab[2,1]*x+para
b[3,1]
:Else
:Return y1(x)
:EndIf
:EndFunc

:DrawFunc g(x)
:EndFor
:@Next loop draws vertical line segments
  at each subdivision
:For k,1,n-1,2
:For i,0,2
:Line xlist[k+i],0,xlist[k+i],ylist[k+i]

:EndFor
:EndFor
:EndPrgm
```

The three "randMat" lines
simply create three matrices for future use.

xp is the 3-by-3 matrix that will later be inverted

The user, when running this program, will
 enter a, b, and n

$xlist$ is a collection of $n + 1$ x-values for each
 subdivision and $ylist$ is the set of the
 corresponding y values for each subdivision

yp is a column vector of 3 y-values for the 3
 points on the parabola
$parab$ is the column vector of the 3 coefficients
 of each parabola

The function $g(x)$ is the function whose graph is
 a parabolic arc for each set of three $xlist$ values.

The parabola is only defined for its 3-point interval.
 The DrawFunc command is used to graph it.

This loop draws the vertical boundaries for each
 $xlist$ value.

FIGURE 12.2.

A listing for graphing the parabolic arcs that form the basis for the calculation of Simpson's Rule.

Upon running the program, the first visible occurrence is that the user is prompted for the values of a, b, and n. These values are input by the user on the "Prgm I/O" screen. Figure 12.3 shows an example with $a = -2$, $b = 2$, and $n = 4$.

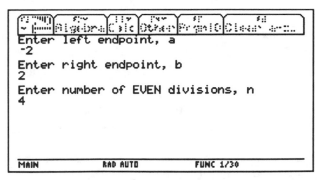

FIGURE 12.3.

Input occurs on the I/O screen. The HOME screen and I/O screen are toggled by the [F5] key.

With $n = 4$, we have two parabolic arcs and they are graphed (along with the function) on the left side of Figure 12.4. After we ran the program with $n = 8$, we got the four parabolic arcs (that are barely distinguishable from the function) on the right side of Figure 12.4 below.

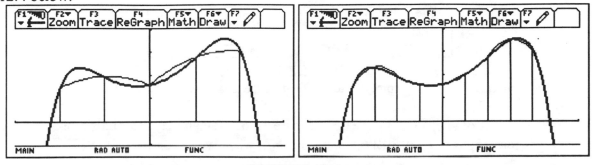

FIGURE 12.4.

Two parabolic arcs on the left with $n = 4$. The four arcs on the right are graphed after running the program with $n = 8$. Notice how closely the area under the parabolas approach the area under the original function.[18]

Procedure #2. Now let's write a program to approximate the area by Simpson's Rule:

$$\int_a^b f(x)\,dx \approx \frac{h}{3}[f(x_0) + 4f(x_1) + 2f(x_2) + 4f(x_3) + \cdots + 2f(x_{n-2}) + 4f(x_{n-1}) + f(x_n)]$$

---

[18]It may not be easy to tell that the arcs are actually part of a parabola. Perhaps the reader will be better convinced with this graph which shows more of the parabolas for the case $n = 4$:

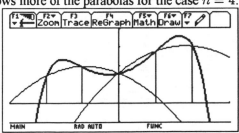

We perform this calculation with the program simpson2(). Because we modularized these programs here, simpson2() is a separate program but this computational segment could be made part of simpson1() just as easily. Note that simpson1() must be run before simpson2() so that the values of a, b, n, and h as well as *ylist* are created.

The "If" statement makes use of a condition check to see if the index of the term is even. If it is even, it is multiplied by 4 and if it is odd it is multiplied by 2. The first and last terms, indexed by 1 and $n + 1$, respectively, are multiplied by 1 (not 4 or 2). The terms are added up and stored in the location called simpsum. Finally, simpsum must be multiplied by the value of $h/3$; that quantity is stored in location finalsum and is displayed in exact and approximate modes. The listing for simpson2() appears in Figure 12.5.

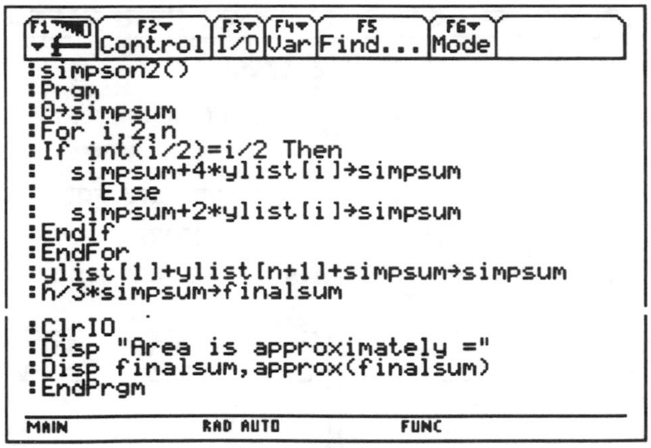

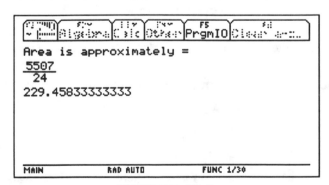

FIGURE 12.5.
The listing for simpson2(), a program to approximate the area under the curve $y1(x) = f(x)$.

The output of simpson2() appears in Figure 12.6. Remember, first run simpson1(). In order to get the output in Figure 12.6, we used $n = 8$ when we ran simpson1().

FIGURE 12.6.
$\int_{-2}^{2}(-x^6 + 20x^2 + 10x + 40)\,dx \approx 229.458$ (with $n = 8$).[19]

---

[19]Of course, this particular example has an *exact* solution: $\int_{-2}^{2}(-x^6 + 20x^2 + 10x + 40)\,dx = \frac{4832}{21} = 230.1$. Our approximation (with eight subdivisions) differs from the exact value by less than 0.3%

Procedure #3. Let's choose a different function – one which has no closed-form (that is, a function which has no elementary anti-derivative): $y = e^{-x^2}$ on the interval $0 \leq x \leq 1$. Instead of calculating a Simpson approximation for this function, we shift our focus to examining the error function, $\dfrac{(b-a)^5 M}{180 n^4}$, for Simpson's approximation. Recall that M is the maximum for $\left| f^{(4)}(x) \right|$ on the interval $a \leq x \leq b$.

We can find M either graphically or analytically. We will use the **TI-92** to graph $\left| \dfrac{d^4 y}{dx^4} \right|$. We enter e^{-x^2} for $y1(x)$ in the $Y =$ editor and deselect it. (We don't need to see the graph for now.) For $y2(x)$ we enter abs(d(y1(x),x,4)). The maximum value of the fourth derivative of $f(x) = e^{-x^2}$ on $0 \leq x \leq 1$ is observed to be 12. See Figure 12.7.

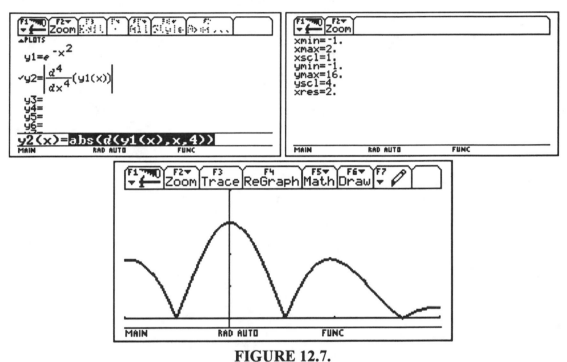

FIGURE 12.7.
The maximum value of $\left| f^{(4)}(x) \right|$ on [0, 1] is 12.

Now, the error in calculating Simpson's Rule is given by $0 \leq S_{err} \leq \dfrac{(b-a)^5 M}{180 n^4}$. With $M = 12$, $a = 0$, and $b = 1$, we can find the value of n that will ensure us having a given degree of precision. Suppose we wanted to find out how many subdivisions, n, will guarantee that the error in our area approximation will be less than 0.0001. The **TI-92** "Solve(" command is used to find the value of n after the values of a, b, and M are stored. See Figure 12.8 below.

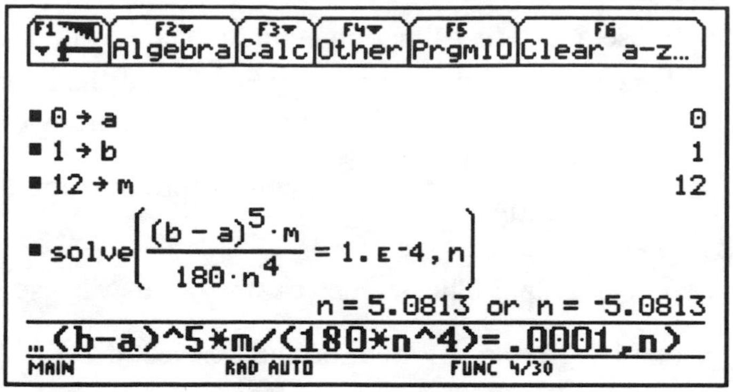

FIGURE 12.8.

The value of n subdivisions that insures the Simpson approximation to be less than 0.0001 is $n = 6$.
(Note: n **must** be even!)

Exercises

In questions 1 through 6, use Simpson's Rule to estimate the integral over the given number of subintervals, n. Also sketch the function and the parabolic arcs as we did in Procedure 1.

1. $\displaystyle\int_1^2 \sqrt{x^3 - 1}\, dx \quad n = 8$

2. $\displaystyle\int_1^2 \sqrt{x^3 - 1}\, dx \quad n = 4$

3. $\displaystyle\int_1^8 \frac{3}{(x+1)^2}\, dx \quad n = 12$

4. $\displaystyle\int_1^{10} \frac{5}{x}\, dx \quad n = 4$

5. $\displaystyle\int_{-4}^4 e^{-x^2}\, dx \quad n = 8$

6. $\displaystyle\int_0^{\sqrt{\pi}} \sin\left(x^2\right) dx \quad n = 8$

7. In questions 3 and 4, the *exact* value of the integral can be found. Use traditional methods to evaluate the definite integrals and compare with the Simpson's Rule approximation found in those exercises.

_____

_____

8. a) Determine the value of n needed using Simpson's Rule to compute $\displaystyle\int_1^2 \frac{1}{x^2}\, dx$ so that the error is less than 0.002.

_____

_____

b) Determine the value of n needed using Simpson's Rule to compute $\displaystyle\int_1^2 \frac{1}{x^2}\, dx$ so that the error is less than 0.00001.

_____

_____

9. Determine $\displaystyle\int_0^1 e^{x^2}\,dx$ to six decimal place accuracy using Simpson's Rule. What value of n is necessary?

_____

_____

10. Considering the fact that Simpson's Rule relies on approximating area with parabolic arcs, it should not be surprising that for quadratic functions, there is no error in Simpson's Rule. What happens if the function is linear, say $y = 2x + 3$?

_____

_____

11. Notice the expression $M = \max\bigl(\bigl|f^{(4)}(x)\bigr|\bigr)$ on the interval $a \le x \le b$ is zero if $f(x)$ is a cubic polynomial and hence, Simpson's approximation is *exact*. How can you justify this? HINT: Examine the function $f(x) = -x^3 + 5x + 2$ on the interval $[0, 2]$. Plot the function along with 3 parabolic arcs $(n = 6)$.

_____

_____

PROGRAMMING THE TI-92

I. Introduction

In the words of Donald Knuth:

> "The process of preparing programs...is especially attractive, not only because it can be economically and scientifically rewarding, but also because it can be an aesthetic experience much like composing poetry or music."[1]

In addition, of course, it can be a frustrating experience which consumes large amounts of time as the size and complexity of your programs increase. While, strictly speaking this is not intended as a "how to program" section, users of the *TI-92*'s programming capabilities are urged to remember that a methodical approach to writing programs will pay handsome dividends. The more thought given to "pre-coding" issues like algorithms and data types will almost always produce programs that are easier to understand and so to maintain.

Having said all that, the programming language supported by the *TI-92* contains most of the features of modern high level computer programming languages including:

- sequential execution of program instructions,
- one-way, two-way, and multi-way selection,
- multiple looping structures,
- subprograms and functions,
- global and local variables,

and a collection of *TI-92* specific features including direct screen addressing, data type conversion commands, and many more. We will demonstrate many of these features.

II. Entering a New Program

To enter a new program, select "7:Program Editor" from the APPS menu and then select "3:New...". The NEW dialogue box appears and for "Type:" select "Program". Then, cursor down to "Variable:" and type *hello1* in the text entry box. Now press ENTER twice. (See Figure 1.)

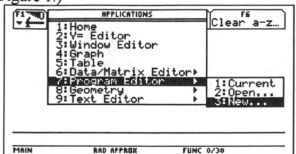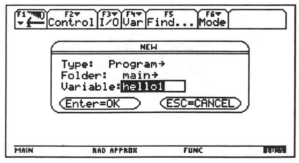

FIGURE 1.
Creating a new program.

---

[1]In the preface of *The Art of Computer Programming, Volume I, Second Edition* (Addison-Wesley: 1973).

We are now in the programming entry screen. The new program entry template appears with the name of our program in place. Let's enter a program that displays a message on the screen when it runs. Keep in mind that every program statement must begin with a colon (:). Just type in the single command **Disp** "Hello World" at the blinking text cursor, then go to the HOME screen, enter the name of our program, hello1(), on the edit line. Observe the result on the program I/O screen. This sequence may be seen in Figures 2 and 3.

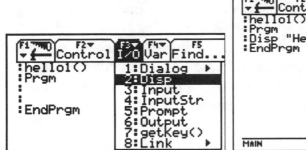

FIGURE 2.

The one line we enter in this program is Disp "Hello World".
The display command, Disp, can either be typed or entered (as we did) via the F3 key.

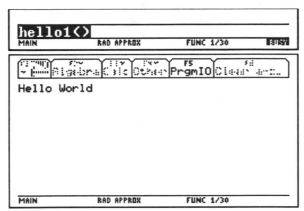

FIGURE 3.

To execute this program, enter its name on the HOME screen (above) and press ENTER. The output of the program is displayed on the *TI-92*'s Prgm I/O screen (bottom).

To re-run the program, return to the HOME screen from the program I/O screen and immediately press ENTER to re-run the program. See Figure 4 below.

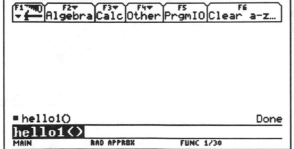

FIGURE 4.

If the program is run a second time, the output is directed to the next available line of the Prgm I/O screen.

We should note several things at this point. First, to run any program, merely enter its name on the edit line of the home screen (or refer to it by name in another program) making sure to include the parentheses () that follow it. Some programs require "start up" information to be passed to them when they are called and it would have to be included within the parentheses on the edit line. Even if the program does not require any initial information, you must still include the parentheses.

Second, notice that the screen on which output for a program is displayed is not the HOME screen. In this case, it is the program I/O screen. There are other possible screens including the GRAPH screen, but not the HOME screen. To switch between the I/O screen and the HOME screen, use the [F5] Prgm I/O key.

Third, notice that when a program is executed, its text output appears on the I/O screen immediately following whatever is currently on the I/O screen. That is, if you want the I/O screen cleared *before* your output is displayed, you must clear it from within your program (or type ClrIO on the HOME screen edit line.) Let's modify home1() so that it clears the I/O screen before displaying its message. We do this by selecting "7:Program" from the [APPS] menu and press "1:Current" to be taken directly to the program entry screen with program hello1() already entered. Place the cursor at the end of the line that says :Prgm and press [ENTER]. Now type the command ClrIO[2], return to the HOME screen and re-run the hello1() program. See Figure 5.

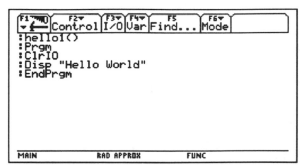 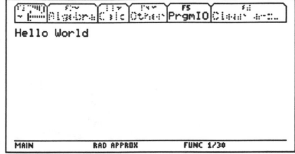

FIGURE 5.
The ClrIO command clears the previous output to the Prgm I/O screen.

For a third variation of this program, return to the HOME screen, and enter the ClrIO command. Then, enter the program editor and start a new program called hello3(). In this program, we will position our "Hello World" message at a randomly selected position on the I/O screen. We will use the **rand()** command to generate 2 random numbers to be used as the pixel coordinates of the starting location for our message. We will also use the **Output** command to perform the actual output. Here's our program:

---

[2]We find it easiest to type a command when we are sure of its syntax (and its case sensitivity). An alternative is to access the command from the CATALOG or, in some cases, from the menu options of the program screen.

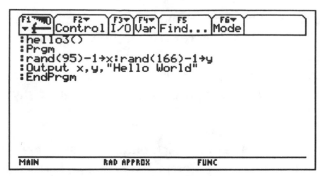

FIGURE 6.

x and y are assigned random numbers and, for convenience, we put both assignment statements on one line separated by a colon.

Before running this program we note that $\rightarrow$ is the symbol generated when the $\boxed{\text{STO} \triangleright}$ key is pressed. Also note that rand(95) generates a random integer between 1 and 95, so rand(95)-1 is a random integer between 0 and 94. This integer is used for the x-coordinate of the screen pixel at which our message begins. Similarly, rand(166)-1 generates the needed y-coordinate, an integer between 0 and 165. We put both commands on the same line (for convenience only) separated by a colon (:). Repeatedly execute this program from the HOME screen to see the I/O screen fill up with our message. You will notice that occasionally, our message overwrites a previous occurrence of our message. This is to be expected since we are generating random starting locations. See Figure 7.

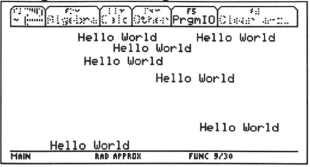

FIGURE 7.

The result of running the program hello3() seven times.[3]

The "magic" numbers 95 and 166 that appear in our program are related to the size of the I/O screen in pixels, 103 by 239, the length of our message, 11 characters including a blank, and the size of the matrix, in pixels, in which each character is composed. Experimentation with variations of this program will help you determine the size of the character matrix as well as the effect of screen location arguments in the Output command that are not between 0 and 103 or between 0 and 238.

---

[3]Your screen may not look exactly like ours if your sequence of random numbers is different.

III. A Function with a Loop

Our next example shows that user-defined functions as well as programs can be written on the *TI-92*. A function is similar to a program but the primary differences between them are that a function *must* return a value and is limited in the programming instructions and built-in functions it may use. With respect to actually using a function, it may be used just like a program, by typing its name and any required input on the edit line of the HOME screen, but unlike a program, it may be used as part of a larger expression as well. In this section, we will write a function to compute the factorial of any non-negative integer.

Select "7:Program Editor" from the APPS menu and then choose "3:New". For "Type:", select "2:Function" and name it fact1. Notice that the function template is used instead of the program template. Now, enter the following function code on the programming screen:

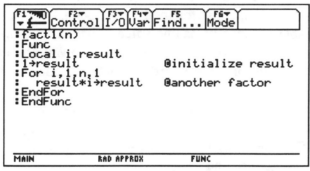

FIGURE 8.
The function definition, *fact1(n)*.

Before actually using this function, we notice several new features. First, when we invoke this function from the edit line of the HOME screen (or as part of some larger expression) we must pass it a non-negative integer upon which to act. This accounts for the presence of the variable n on the function name line, fact1(n). Second, the programming language for the *TI-92* permits variables to be declared prior to their use in a program or function. In this case, we need two (2) variables which have no life outside our function-- thus, we declare them as local variables and they will disappear as soon as our function has performed its calculations.

Notice also the presence of **comments**...text immediately preceded on a line with the © symbol. They are merely documentary remarks intended to help us understand what our code does when we return to it a few days or weeks after we enter it. Finally, notice the use of the looping structure **For...EndFor**. It is one of several repetition constructs that the *TI-92* programming language supports and is designed to repeat a collection of statements a known number of times. Its syntax requires a loop control variable (*i* in our case) followed by the starting and halting values of that variable, followed by an optional incrementing value, all separated by commas. The default for the optional increment value is 1. As with many of the programming statements, the For...EndFor template may be typed in from the keyboard, or it may be obtained from one of the menus on the program edit screen. Since it is a program control structure, we expect to find it as an option under the F2 Control menu. It is there as item "4:For...EndFor". It is wise to obtain it from the menu since all of the "pieces" of the statement are automatically inserted into the program making it unlikely that you, the programmer, will inadvertently leave a piece out. For instance, Disp is available under F3 I/O as option "2:Disp", Output is available under F3 I/O as "6:Output", and Local is available under F4 Var as "3:Local".

It is important to understand how the For...EndFor statement works. The statements between the For and EndFor statements are collectively called the *loop body* and are repeatedly executed as long as the loop control variable has not exceeded its halting value. In practice, when the For statement is encountered, the loop control variable is initialized with the starting value and then compared to its halting value. If it is no larger than its halting value, the loop body is executed, the loop control variable then incremented by the incrementing value, and the termination test performed again. This continues until the value of the loop control variable exceeds its halting value at which time the loop body is skipped and execution resumes at the first executable statement following the EndFor statement. It is a good idea to "hand check" this program in order to become familiar with this structure.

Before continuing, execute the function fact1() several times from the HOME screen edit line, with various arguments, and in various expressions. See Figure 9 for some samples. You can see that fact1() is certainly not "bulletproof". Can you explain its behavior when given a negative or non-integer argument?

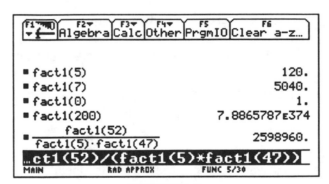

FIGURE 9.
Five variations of invoking the function *fact1* from the HOME screen.

IV. The Return Statement, Two-Way Selection, and Recursion

In function fact1(), the last calculated value, result*i STO ▷ result, is the value returned by the function. But this is not the only way to have a function return a value. Consider the following variation called fact2().

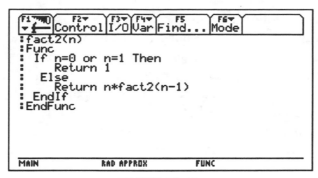

FIGURE 10.
The function called *fact2*.

Note, first, the use of the control structure **If...Then...Else...EndIf** called a *two-way selection statement*. Its template may by typed in or selected from the [F2] Control menu as option "2:If...Then...Else...EndIf". In case fact2() is invoked with an argument of 0 or 1, the *Then* path is followed and a value of 1 is returned and the function terminates. If anything other than 0 or 1 is passed to fact2(), say fact2(7), the Else path is followed and the value 7*fact2(6) is returned and then the function terminates. This is all well and good provided that fact2(6) is calculated properly. Don't worry--it is. This kind of function (or program) that calls itself to calculate a value is called a *recursive function* (or program). The **TI-92** supports recursion--and it is a very powerful feature in a programming language.

Now, run fact2() several times with a variety of arguments and compare its output to that of fact1() for "bad" input. Fact2() calculates no value (instead, it runs out of memory), whereas fact1() produces erroneous output.[4] As you can see, error-checking can be critically important in producing robust, error-free programs.

Before moving on, we should mention that while the use of recursion often leads to a natural, easy-to-visualize program, a non-recursive (iterative) version may be more efficient with respect to time and use of memory.

V. Addressing the Graphics Screen: The Sierpinski Triangle

The beautiful field of fractals leads us to our next example--the *Sierpinski triangle*. A fractal may be thought of as a geometric shape having the property that it can be broken into parts, each of which has (at least approximately) the same shape as the whole. Fractals often have simple mathematical descriptions.

---

[4]*Caution:* When the **TI-92** runs out of memory, it is possible that you will lose all stored programs.

Imagine playing the following game, called the "Chaos Game."[5] You choose any 3 non-collinear points in the plane, thus forming a triangle with vertices A, B, and C. Then, pick some arbitrary starting point, say P1, anywhere in the plane. Now, randomly (using any distribution) select one of the 3 original vertices and locate the point 1/2 way between P1 and the selected vertex. Call this new point P2. Now, repeat the process, randomly selecting a vertex and locating the point 1/2 way between P2 and the selected vertex. Continue the process many times. Can you imagine what the locus of the selected points P1, P2, P3, ..., looks like? Let's write a program to find out.

Enter the new program, *points*, as it appears in Figure 11 using the Program Editor:

FIGURE 11.

The program to generate the Sierpinski Triangle, *points.*

Now run the program several times making sure to include a positive integer argument each time. You should keep in mind that the larger the integer argument, the longer it will take to complete. For example, *point(500)* takes about 1.5 minutes to complete.

Before talking about what "picture" results, let's look at the program code. First note that this program requires a parameter, *n*, passed to it when it is invoked. This positive integer determines how many points are plotted. Next, notice that we make use of 4 local variables--1 for loop control (i), 2 for the point coordinates (x, y), and 1 to hold the number of the randomly chosen vertex (v). The statements **ClrGraph**, **ClrDraw**, and **FnOff** are "housekeeping" chores to ensure that the graph screen is clear and that no Y = defined

---

[5]The Chaos Game was so named by Michael Barnsley in his book, *Fractals Everywhere*, Academic Press, 1989.

function will be drawn as our points are plotted. The setGraph command turns the axes off to present an uncluttered picture.

As we wrote our program we had in mind an imaginary triangle whose vertex coordinates are V1(0,4), V2(6,0), V3(-6,0) which accounts for our choice of WINDOW settings and also the initial values of x and y in our program. You may prefer other choices.

Notice the use of a comment (©) to remind us what this statement is doing. Comments may be inserted anywhere in a program, but typically they are put on a line by themselves or else at the end of a line following some instruction. The © symbol (accessed either by [F2] "9:©" or by typing [2nd] [X]) signals the start of a comment and everything following it on a line is treated as a comment and ignored by the *TI-92*.

The For loop is repeated as many times as the user specifies at invocation of *points()* and after entering the loop body, we generate a random number to determine our next vertex. We then use the multi-way selection structure If...Then...ElseIf which allows us to take one of 3 courses of action, depending upon which vertex we have randomly selected. The action is merely to calculate the coordinates of the point midway between our last point and the selected vertex. After the appropriate selection has been made and the coordinates of the new point calculated, we "plot" that point using **PtOn** to turn on the screen pixel nearest to the coordinates of the point. We emphasize that any 3 (noncollinear) points would do for our starting triangle as well as any initial point--we just chose an isosceles triangle, centered in the middle of the screen, with a starting point in the middle of the triangle.

As you probably have seen by now, the resulting "figure" is a triangle with certain triangular sections missing. This is called Sierpinski's triangle and it can also be constructed in a straight forward geometric manner. Begin, as usual, by selecting any triangle. Next, connect the midpoints of each side to form a new triangle which you then remove, leaving 3 smaller triangles. Repeating this process on the each of the 3 smaller triangles, leaves 9 triangles. Repeated iterations produce more and more triangles--each a self-similar copy of the whole. Particularly interesting is the fact that if the original triangle has area 1, say, the Sierpinski triangle has zero area, but infinite circumference! Figure 12 shows the graphics screen resulting when we ran *points(800)*.

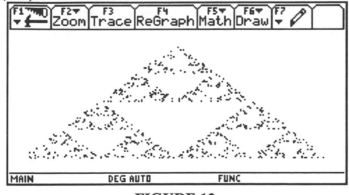

FIGURE 12.
The Sierpinski Triangle created from running *points(800)*.

VI. A Guessing Game Using a While Loop and Pop-Up Menus

We now write a program that uses a variety of *TI-92* programming tools including a While Loop, PopUp menus, Input and Pause statements, and a Boolean variable.

The game we intend to program goes like this: the player first chooses integer upper and lower bounds, and then thinks of an integer between them, but does not reveal it to the program. By asking a series of Yes/No questions, the program attempts to guess the player's number. Figure 13 displays the code for program *bisearch* (read down the left column of code and then read down the right column):

```
F1        F2      F3  F4    F5      F6
   Control  I/O Var Find... Mode

:bisearch()
:Prgm
:ClrIO
:Local low,high,found
:Local numguess,middle,guess,reply
:Disp "To play this game, you first"
:Disp "need to enter integer lower and"
:Disp "upper bounds.  Then think of an"
:Disp "integer between those bounds."
:Disp "This program will then guess"
:Disp "your number--with your help."
:Loop
:  Input "Enter lower bound",low
:  Input "Enter upper bound",high
:  If low<high Then
:    Exit
:  Else
:    Disp "Lower must be less than upper."

:    Disp "Please re-enter."
:    Disp ""       @ a blank output line
:  EndIf
:EndLoop

:Disp "Think of an integer between bounds."
:Pause "Press ENTER when ready ..."
:false→found    @ Boolean for while loop
:0→numguess     @ initialize counter
:While not found and low≤high
:  int((low+high)/2)→middle
:  middle→guess
:  numguess+1→numguess
:  ClrIO
:  Disp "Is this your number?",guess
```

```
:  PopUp {"Yes","Guess is too high","Guess is too low"},reply
:  If reply=1 Then
:    true→found        @Found it!
:  ElseIf reply=2 Then
:    middle-1→high
:  ElseIf reply=3 Then
:    middle+1→low
:  EndIf
:EndWhile
:If found Then
:  Disp ""           @Found is true

:  Disp "I knew I could do it!"
:  Disp "And it only took",numguess
:  If numguess=1 Then
:    Disp "guess."
:  Else
:    Disp "guesses."
:  EndIf
:Else
:  Disp "Hey...you're not playing fair."
:EndIf
:EndPrgm
```

FIGURE 13.
The code for the guessing game.

Before we discuss the program, you may wish to enter bisearch() in the Program Editor and run it several times with a variety of bounds to get a "feel" for what it does and maybe even how it does it.

Now, let's talk about the program. First, notice the generous use of local variables. They not only make good programming sense (since none of them has any reason to exist outside of the program), but they also make more efficient use of the *TI-92*'s memory than global variables (those not declared as **Local**) which remain in memory even after program termination.

Following some game instructions to the user of the program, we make use of the **Loop...EndLoop** structure to get the lower and upper bounds into the program as well as checking them for order. The **Loop...EndLoop** construct is, in fact, an infinite loop structure and must be used with care. The programmer is responsible for including in the loop body a mechanism to halt the loop. If no "escape" is provided, the only way to exit the loop is to abort the program by pressing the ON key during execution. The purpose of the loop in our program is to make sure the player enters a lower bound that is actually less than the upper bound. If so, the **Exit** statement breaks out of the loop and the program resumes execution at the first executable statement following the **EndLoop** statement.

The **Input** statement provides a convenient way to bring data into a program while prompting the user with an optional message. The **Pause** statement is often used when the screen is full of text that may take a moment to read, or when the user needs time to consider what action to take.

You may have noticed what looks like a variable's value being stored in another variable--**false-->found**. Actually, we are storing a Boolean value, *false*, in a Boolean variable, found. The reason for this is that it is one of 2 conditions that will terminate our search for the player's number...namely, finding it. We will explain the other shortly.

After initializing the number of guesses we've made so far to zero, we get to the real heart of the program...the **While...EndWhile** loop. Such loops provide a means of repeatedly executing a collection of statements subject to a Boolean condition. Unlike the **For** loop, which is executed a known number of times, the **While** loop is repeatedly executed as long as its termination condition is not met. In our case, as long as we have not yet found the number and low is still no bigger than high, we execute the statements in the loop body. Of course, unless we take steps within the loop body to alter the termination conditions, we will never leave the loop.

So how do we alter the termination conditions? Either by guessing the correct number, in which case we set the Boolean variable found to true (making *not found* have the value false) or by making low > high. To understand how this may happen, notice that all we do is constantly make our guess the number half way between low and high. If our guess is too high, we adjust variable high to become one less than our guess and then calculate a new guess which is lower. If our guess is too low, we adjust the variable low to become one more than our guess and then calculate a new guess which is higher. Is it any wonder, then, why this kind of search is called a *binary search*?

In order to help the player enter only the appropriate choices, we provide a **PopUp** menu from which to select one. That way, we are sure of exactly what the user will be responding and need only make provisions in our program for those options. The syntax should be pretty clear...provide a string (anything enclosed in double quotes) for each choice on the menu and a variable to hold the number of the choice selected by the player when the program runs. We then include the appropriate **If...Then...ElseIf...EndIf** structure to account for each possibility.

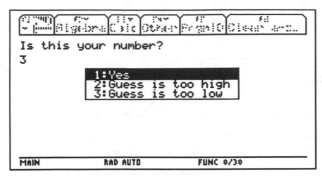

FIGURE 14.
A partial run of the program bisearch() showing the use of the **PopUp** command.

Finally, it is worth saying something about the maximum number of guesses needed by a binary search program to either locate the number or determine that it's not within bounds. The repeated division of our interval of possibilities in half to make our guesses ensures that we never need more than $1+\text{int}(\log_2(\text{high} - \text{low}))$. For example, if you pick the interval from 1 to 10,000, $1+\text{int}(\log_2(9,999)) = 1 + 13 = 14$. So the program would never need more than 14 guesses to locate the player's number or conclude the player didn't play by the rules. A few judicious choices for bounds should make this clear.

VII. Graphing the Mean Value Theorem for Derivatives

As you may know, the graph of a function often reveals many of the function's properties including continuity, differentiability, and concavity. It's no wonder, then, that many theorems in calculus have a strong geometric "flavor". Among them is one of the most useful results in all of calculus--*The Mean Value Theorem for Derivatives* (MVTD).

The program mvtd(), seen below in Figure 15, was written to create a graphical view of this theorem. In essence, the MVTD says that if a function, $y = f(x)$, is "well-behaved" on the closed interval, $[a, b]$, then at some point on the curve between $x = a$ and $x = b$, the slope of the function is the same as the slope of the chord connecting the points $(a, f(a))$ and $(b, f(b))$. When we say *the slope of the curve*, we really mean the *slope of the tangent line to the curve at that point*. In the language of algebra, this says that there is at least one value of x, call it x_0, in the open interval (a, b) such that

$$\text{the slope of the curve at } x_0 = \frac{f(b) - f(a)}{b - a} \tag{1}$$

The right-hand side of the equation above is just the slope of the chord joining $(a, f(a))$ and $(b, f(b))$. In the first semester of any calculus course, we learn that the left-hand side of the equation above can be found by evaluating the first derivative of $f(x)$ at x_0...that is, $f'(x_0)$. The program mvtd() graphically illustrates that the tangent line to $y = f(x)$ at each of the solutions to the equation is parallel to the chord.

Here is our program:

```
F1---0  F2-   F3-  F4-  F5     F6-
--+--Control I/O Var Find... Mode
:mvtd()
:Prgm
:Local a,b,xlist,i
:FnOff
:FnOn 1
:ClrIO
:Disp "This program uses the current WIN
DOW"
:Disp "settings as well as the current v
alue"
:Disp "of y1(x) which is"
:Disp y1(x)
:Disp "Press ENTER to continue"
:Pause "Press ON key to ESCAPE..."
:Disp
:Input "Enter left endpoint",a
:Input "Enter right endpoint",b
:exp▶list(solve(d(y1(x),x)=(y1(b)-y1(a))
/(b-a),x)|x>a and x<b,x)→xlist
:ClrIO
:Disp "The set of x-values satisfying"
:Disp "the conclusion of MVTD is"
:Disp xlist
:Disp "Press ENTER to see graph of y1(x)
"
:Disp "with chord drawn from a to b"
:Disp
:Disp "Then press ENTER to see tangent(s
)"
:Pause "drawn at solution points."
:Style 1,"thick"
:DispG
:Line a,y1(a),b,y1(b)
:Pause
:For i,1,dim(xlist)
:   LineTan y1(x),xlist[i]
:EndFor
:Trace            @leave in Trace mode
:Style 1,"line"
:EndPrgm
MAIN          RAD AUTO        FUNC
```

FIGURE 15.
A listing of mvtd().

As you can see from reading the code, this program expects the user to have already entered the target function as y1(x) in the Y = Editor, and graphed it over the interval of interest using some appropriate viewing WINDOW, before running this program. As a word of advice, you may wish to avoid the periodic trigonometric functions because of the way the solve() function works with them.

Run the program several times on functions of your choice. Just remember, by a well-behaved function in this context, we mean one that is continuous at every point in the closed interval $[a, b]$ (has no holes, jumps or vertical asymptotes), and is differentiable at every point in the open interval (a, b) (roughly speaking, has no sharp corners). For example, you may wish to try $y1(x) = x^{1/3}$ on the closed interval $[-8, 1]$. First, enter the function in the Y = Editor and graph it in some WINDOW in which you can see the function over this domain...say xmin = -10, xmax = 2, ymin = -4 and ymax = 2. When prompted for the left endpoint enter -8 (not the WINDOW value of xmin) and enter 1 when prompted for the right endpoint (not the WINDOW value of xmax).

Try the program again with $y1(x) = -x^3 + 4x + 1$ on $[-2, 3]$. You may find WINDOW settings of xmin $= -6.5$, xmax $= 6.5$, ymin $= -15$, and ymax $= 10$ are appropriate. We include, in Figure 16, the screens that make up the run of *mvtd()* with this function.

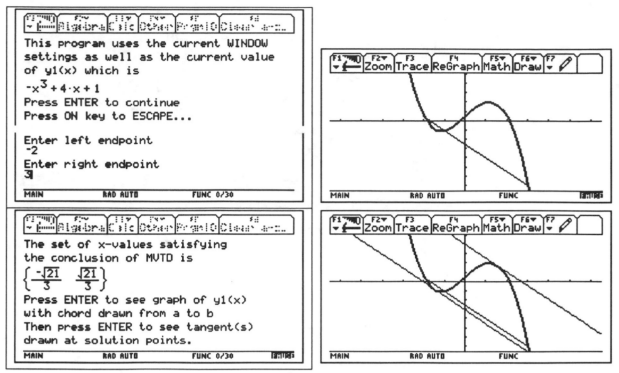

FIGURE 16.
A run of mvtd() using $y1(x) = -x^3 + 4x + 1$ with $a = -2$ and $b = 3$.

In many examples including the one displayed in Figure 16, you may have noticed that there is more than just one point in the designated interval for which the conclusion of the MVTD was true. This does not contradict the theorem which says only that *at least* one such point exists.

You probably have a good idea of how the program works by now, so let's discuss some of its new features. **FnOff** turns off all Y = defined functions which may be selected. We do this so that our graph isn't cluttered with unwanted functions. **FnOn 1** turns y1(x) back on so that it will eventually be graphed. We include a message reminding the user that the target function must reside in y1(x) and also remind the user that to exit the program at that time, press the ON key, followed by the ESC key. There are several **Pause** statements to allow the user to consider decisions or to reflect on the output.

An important part of this program is the solution of equation (1) above. We use the **solve()** command with a restricted domain. We are only interested in solutions strictly between a and b--nowhere else. This accounts for the use of the **with operator** ($|$). The **exp ▷ list()** function which precedes the solve() command puts the solution(s) in a list instead of displaying them in the "x=1 or x=2" form. When they are in a list, they can be used later to control a loop that draws the tangents.

The **Style 1, "thick"** command makes sure that when y1(x) is graphed with the **DispG** command, it will be drawn in the "thick" style. The *TI-92* provides full access to graphing modes and styles and we take advantage of that in mvtd(). Following the graphing of y1(x) by DispG, the **Line a,y1(a),b,y1(b)** command draws the line segment (chord) between the left and right endpoints on the graph and then the program **Pauses** until the user presses ENTER to continue.

At this point, the **For...EndFor** structure draws a tangent line corresponding to each solution(s) in xlist using the number of elements in xlist (**dim(xlist)**) as the number of times to enter the loop body. Note that this even works when there aren't any solutions, in which case dim(xlist)=0 and the loop is never entered. Of course, if we use only functions for y1(x) that satisfy the hypotheses of the MVTD, there will always be at least one solution, so at least one tangent line will be drawn.

The statement that actually draws the tangent lines is the **LineTan** statement which needs the expression that defines the function, y1(x), and an x-value (or list of x-values) for the points on the curve through which a tangent is to be drawn. Finally, the **Trace** command leaves the Graph screen in Trace mode for exploration and **Style 1, "line"** leaves y1(x) in "line" style.

It is worth noting that, once again, this program is not "bulletproof". There are several ways that you can make the program "crash" or perform peculiarly if that is your intention. But, like most things in life, programming involves tradeoffs...in this case, ease of use and coding, versus program complexity and memory.

VIII. Creating Toolbars to be Used with the CUSTOM Key

As you use the *TI-92,* you may notice that some symbols or commands are difficult to find. By now, you may feel comfortable enough with the syntax of the symbols or commands to type them directly from the keyboard. But if you would like to create a customized toolbar that serves as a convenient way of accessing these items, then you may want to consider the program below.

The program *custprep()* prepares a custom menu system. Once it is run, a set of your customized titles will appear across the top of the screen by pressing 2ⁿᵈ CUSTOM. Here is the program:

```
F1▼┐┌F2▼┐┌F3▼┐┌F4▼┐┌ F5 ┐┌F6▼┐┌───────┐
▼←─┘│Control│I/O│Var│Find...│Mode│
:custprep()
:Prgm
:Custom
:Title    "Symbols"
:Item     "!"
:Item     "°"
:Item     "r"
:Item     "%"
:Title    "Relations"
:Item     "≠"
:Item     "≤"
:Item     "≥"
:Title    "Vectors"
:Item     "crossP("
:Item     "dotP("
:Title    "Conversions"
:Item     "exp▶list("
:Item     "list▶mat("
:Item     "mat▶list("
:EndCustm
:EndPrgm
────────────────────────────────────────
MAIN        RAD AUTO         FUNC
```

FIGURE 17.

A listing of the program *custprep()*.

The program contains a block of statements between "**Custom**" and "**EndCustm**". Each of those statements is either a "**Title**" (of a pull-down menu) or "**Item**" (entry in a pull-down menu) statement. We have included as items some of the things that we find hard to find, yet we need them often.

When the program is run, the custom menu becomes accessible. To activate it, press 2nd CUSTOM. Observe the top of the screen in Figure 18 below.

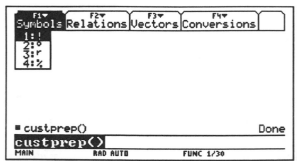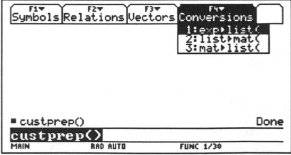

FIGURE 18.

After running *custprep()*, press 2nd CUSTOM. The four *Titles* appear across the top of the HOME screen until 2nd CUSTOM is pressed again. We displayed only the items for the first and fourth titles.

The customized items are then easily accessed. Two examples appear in Figures 19 and 20. Figure 19 shows the calculation of the sine of 45° even though we are in radian mode. Figure 20 shows a selection from the F3 Vectors menu.

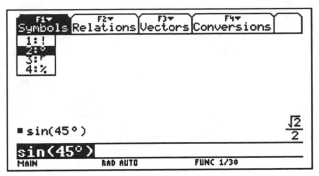

FIGURE 19.

Accessing the item "degree" from the customized menu. After sin(45 is entered, press [F1] then [2] then [)].

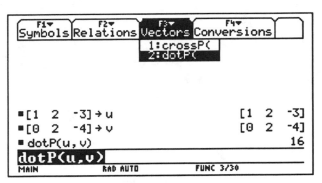

FIGURE 20.

Under [F3], we access the item "dotP(" to find the dot product of vectors **u** and **v**.

NOTES

Index

Greek Letter	Keystrokes
α	[2nd] [G] a
β	[2nd] [G] b
δ	[2nd] [G] d
Δ	[2nd] [G] [SHIFT] [D]
ϵ	[2nd] [G] e
ϕ	[2nd] [G] f
γ	[2nd] [G] g
Γ	[2nd] [G] [SHIFT] [G]
λ	[2nd] [G] l
μ	[2nd] [G] m
π	[2nd] [G] p
Π	[2nd] [G] [SHIFT] [P]
ρ	[2nd] [G] r
σ	[2nd] [G] s
Σ	[2nd] [G] [SHIFT] [S]
τ	[2nd] [G] t
ω	[2nd] [G] w
Ω	[2nd] [G] [SHIFT] [W]
ξ	[2nd] [G] x
ψ	[2nd] [G] y
ζ	[2nd] [G] z

Keystrokes for the 21 available Greek letters (5 of which are capital letters).

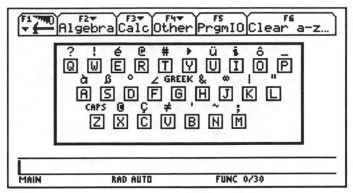

The symbol keyboard template.

A question mark, ?, for example is obtained by pressing 2ⁿᵈ Q

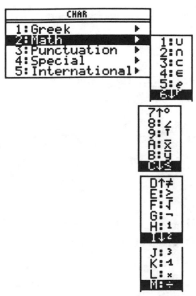

The twenty-two available mathematical symbols found by accessing the 2ⁿᵈ CHAR key.

Some TI-92 Tips

- When you're in EXACT or AUTO modes, pressing ◇ ENTER will approximate a result.

- Two or more commands can be entered on one line by separating them with a colon.

- To rename programs or variables use 2nd VAR-LINK and choose F1 "Manage."

- To return to the Edit line from anywhere in the HOME screen history, simply press ESC.

- 2nd APPS is used to jump between split screens.

- Select menu items by number or letter -- it is quicker than scrolling to your choice and then pressing ENTER.

- Did you forget where (in which menu) a particular command is? If you know its syntax, just type it from the keyboard. Otherwise, you may find it handily in the CATALOG. (Press a letter from within the catalog, and you jump to that part of the alphabet!)

- Lines can be copied and pasted from one screen to another or within one screen. ◇C copies and ◇V pastes. Also, ◇X cuts (copies and deletes) highlighted text.

- In the 2nd CATALOG, the syntax for highlighted commands appears in very small type at the bottom of the screen.

- Press the 2nd key before cursor down (or cursor up) to get to the bottom (or top) of a long menu or list.

- Access the special symbols screen by pressing ◇K.

- To leave the Program I/O screen, either press ESC or F5 (which toggles between the HOME and I/O screens).

- To cancel a program's execution, press ON then ESC.

- To stop a function from being graphed, press ON; to pause it, press ENTER and then ENTER again to resume the graphing.

- If a is a matrix, a^{-1} represents its inverse.

- In programs, results of calculations are not automatically displayed. Use either the Disp or Output commands to display the results on the I/O screen or store them into a variable using the STO▷ key.

- To run a program whose name is *progname*, for example, you must enter progname() from the HOME screen. (You must include the parentheses even if no parameters are being passed to the program.)

- Functions cannot call programs.

- In the Geometry screen, press ESC at any time to get the pointer tool.

- In the Geometry screen, editing numerical data requires ESC (not ENTER) to terminate input.

- In the Geometry screen, the present tool stays active until a new tool is chosen. (Exception: the F4 macro tool).

- In the Geometry screen, a polygon created by segments, is not recognized as a polygon. Thus, the area and perimeter tools won't work -- you must use the calculate tool instead.